Environmental Values

We live in a world confronted by mounting environmental problems. We read of increasing global deforestation and desertification, loss of species diversity, pollution and global warming. In everyday life people mourn the loss of valued landscapes and urban spaces. Underlying these problems are conflicting priorities and values. Yet dominant approaches to policy making seem ill-equipped to capture the various ways in which the environment matters to us.

Environmental Values introduces readers to these issues by presenting, and then challenging, two dominant approaches to environmental decision making, one from environmental economics, the other from environmental philosophy. The authors present a sustained case for questioning the underlying ethical theories of both of these traditions. They defend a pluralistic alternative rooted in the rich everyday relations of humans to the environments they inhabit, providing a path for integrating human needs with environmental protection through an understanding of the narrative and history of particular places. The book examines the implications of this approach for policy issues such as biodiversity, conservation and sustainability.

The book is written in a clear and accessible style for an interdisciplinary audience. It will be ideal for student use in environmental courses in geography, economics, philosophy, politics and sociology. It will also be of wider interest to policy makers and the concerned general reader.

John O'Neill is Professor of Political Economy at The University of Manchester.

Alan Holland is Emeritus Professor of Applied Philosophy at Lancaster University.

Andrew Light is Assistant Professor of Philosophy and Public Affairs at the University of Washington, Seattle.

Routledge Introductions to Environment Series
Published and Forthcoming Titles

Routledge Introductions to Environment Series

Environmental Values

John O'Neill, Alan Holland and Andrew Light

Routledge
Taylor & Francis Group

LONDON AND NEW YORK

First published 2008
by Routledge
2 Park Square, Milton Park, Abingdon, Oxon OX14 4RN

Simultaneously published in the USA and Canada
by Routledge
270 Madison Avenue, New York, NY 10016

Reprinted 2008

Transferred to Digital Printing 2009

Routledge is an imprint of the Taylor & Francis Group, an informa business

© 2008 John O'Neill, Alan Holland and Andrew Light

Typeset in Times New Roman by Florence Production Ltd, Stoodleigh, Devon
Printed and bound in Great Britain by TJI Digital, Padstow, Cornwall

British Library Cataloguing in Publication Data
A catalogue record for this book is available from the British Library

Library of Congress Cataloging in Publication Data
O'Neill, John, 1956–
  Environmental values / John O'Neill, Alan Holland, and Andrew Light.
    p. cm – (Routledge introductions to environment)
  Includes bibliographical references and index.
  1. Environmental ethics. 2. Environmental policy.
  I. Holland, Alan, 1939– II. Light, Andrew, 1966– III. Title.
  GE42.O54 2007
  179'.1–dc22   2006023830

ISBN10: 0–415–14508–2 (hbk)
ISBN10: 0–415–14509–0 (pbk)
ISBN10: 0–203–63952–9 (ebk)

ISBN13: 978–0–415–14509–1 (pbk)
ISBN13: 978–0–415–14508–4 (hbk)
ISBN13: 978–0–203–63952–8 (ebk)

Contents

List of Figures

Acknowledgements

Parts of chapters 4, 10 and 11 draw on material in John O'Neill *Markets, Deliberation and the Environment* (London: Routledge, 2007), John O'Neill 'Sustainability: Ethics, Knowledge and Politics' in J. O'Neill, I. Bateman and R. K. Turner (eds) *Environmental Ethics and Philosophy* (Aldershot: Edward Elgar, 2001), and John O'Neill and Alan Holland 'Two Approaches to Biodiversity Value' in D. Posey (ed.) *Cultural and Spiritual Values of Biodiversity* (London: UNEP, 1999). Chapter 1 draws on material in Alan Holland, Martin O'Connor and John O'Neill *Costing Environmental Damage* (Report for the European Parliament, Directorate General for Research, STOA, Brussels, 1996). Chapter 7 draws on material in John O'Neill 'Meta-ethics' in D. Jamieson (ed.) *Blackwell Companion to Environmental Philosophy* (Oxford: Blackwell, 2001). Chapter 9 draws on material in John O'Neill and Alan Holland 'Yew Trees, Butterflies, Rotting Boots and Washing Lines' in A. Light and A. de-Shalit (eds) *Reasoning in Environmental Practice: Philosophy and Politics in Global Ecological Perspective* (Cambridge, Mass.: MIT Press, 2003). Parts of chapters 6 and 10 draw on material in Andrew Light 'Contemporary Environmental Ethics: From Metaethics to Public Philosophy', *Metaphilosophy* 2002, 33: 426–449. Some of the ideas for this book began their life in work towards teaching material produced for Wye College. Others involved in producing that material were John Benson and Jeremy Roxbee Cox and we would like to record our thanks for their conversations and fruitful collaboration.

1 Values and the environment

Environments and values

This is a book about the environment and about values. However, at the outset it is important to register two seemingly perverse points about this topic:

1. There is no such *thing* as the environment. *The* environment – singular – does not exist. In its basic sense to talk of the environment is to talk of the environs or surroundings *of* some person, being or community. To talk of the environment is always elliptical: it is always possible to ask 'whose environment?' In practice talk of the environment is at best a shorthand way of referring to a variety of places, processes and objects that matter, for good or bad, to particular beings and communities: forests, cities, seas, weather, houses, marshlands, beaches, mountains, quarries, gardens, roads and rubbish heaps.
2. There are no such *things* as values. There are rather the various ways in which individuals, processes and places matter, our various modes of relating to them, and the various considerations that enter into our deliberations about action. Environments – plural – and their constituents, good and bad, matter to us in different ways. First, we live *from* them – they are the means to our existence. Second, we live *in* them – they are our homes and familiar places in which everyday life takes place and draws its meaning, and in which personal and social histories are embodied. Third, we live *with* them – our lives take place against the backdrop of a natural world that existed before us and will continue to exist beyond the life of the last human, a world that we enter and for which awe and wonder are appropriate responses. These different relations to the world all bring with them different sources of environmental concern.

Living from the world

We live from the world: we mine its resources; cultivate and harvest its fruits; shape the contours of the land for human habitation, roads, minerals and agriculture; dredge rivers for transport. And all these activities are subject to the action of the natural world: flood, drought, hurricane, earthquake and landslide can be a source of ruined endeavour and human sorrow. Human life, health and economic productivity are dependent upon the natural and cultivated ecological systems in which we live – on their capacity to assimilate the wastes of economic activity and to provide its raw materials. The damage that economic activity does to these capacities, accordingly, is a major source of increased environmental concern. The effects of pollution directly on the health and life chances of citizens and on the productivity of agriculture, forests, and fisheries, the depletion of natural resources – of fishery stocks, mineral reserves and drinkable water – have all served to highlight the environmental problem in an immediate way both to the general citizenry and to policy makers. At the same time there is growing evidence of global risks to the ecological systems upon which human life depends, such as the depletion of the ozone layer and accelerating rate of climate change – including the threat of global warming. For many people at present these may have little immediate impact – and that mostly localised – but this could soon change. Their implications for human welfare more generally are subject to scientific uncertainty, though here again there is a growing consensus that the effects are unlikely to be benign. Taken together, these sources of concern have given rise to the perception of a global environmental crisis that is in part fuelled by the very invisibility and uncertainty of the risks involved.

Living in the world

We live in the world. The environment is not just a physical precondition for human life and productive activity, it is where humans (and other species) lead their lives. Environments matter to us for social, aesthetic and cultural reasons. Some of this dimension often comes under the heading of 'recreation value' in economic texts, and for some part of the role that the environment plays in human life the term is a quite proper one: it catches the way in which forests, beaches, mountains and rivers are places in which social and individual recreational activities – of walking, fishing, climbing, swimming, of family picnics and play – take place. With some stretching of the term, elements of the aesthetic appreciation of landscape might also come under the heading of 'recreation'. Concerns about quality of bathing water, the loss of recreational fish stocks, and the visual impact of quarries or open-cast mines, in part reflect this value. However, the term 'recreation' can be misleading in the sense that it suggests a view of the natural environment as merely a playground or spectacle, which might have substitutes in a local gym, or art gallery, whereas the places in question might have a different and more central part in the social identities of individuals and communities. Particular places matter to both individuals and communities in virtue of embodying their history and cultural identities. The loss of aesthetically and culturally significant landscapes or the despoliation of particular areas matters in virtue of this fact. Thus, for example, the public significance attached to the damage to forests and lakes in Scandinavia and Germany reflects their cultural as much as their economic importance. This social and cultural dimension also has a more local aspect, for example in the importance that local communities place on the 'ordinary' places in or near which they live – a pond or copse of woods – places that from the economic or biological point of view have limited significance (Clifford and King 1993). The cultural dimension is also realised in issues concerning the quality of the urban environment: in the kinds of social life that different urban environments make possible, the effects of the car not only on the quiet of the city, but also on the capacity of individuals to meet in public spaces: in the heritage the built environment embodies and the sources of cultural identity it provides, and hence concern with the effects of pollution and urban development on that environment.

Living with the world

We live with the world: the physical and natural worlds have histories that stretch out before humans emerged and have futures that will continue beyond the disappearance of the human species. This fact is one to which environmentalists often make appeal. Correspondingly a source of growth in environmental concern manifest in the nature conservation movement has been the steady accumulation of data provided by the life sciences. The loss of biodiversity, the disappearance of particular habitats and the extinction, local and global, of particular species of flora and fauna have all become increasingly central to public debate and policy making. While these issues sometimes have an economic dimension – for example, it may be that there are herbs and medicines that will be lost, or resources that will disappear, and a cultural and aesthetic dimension (the loss of significant forests in Germany had both) – the concern for the environment is not reducible to these. Often such indirect justifications for concern look weak: it is doubtful that human life chances or economic productivity will be much affected by the loss of the blue whale or the red raft spider, or the disappearance of marshland. And significant biological sites can be quite ugly. The supermarket trolley rule of thumb for assessing the biological significance of ponds – the more supermarket trolleys the greater the significance – may not be an exceptionless guide, but it points to the divergence between the beauty or cultural significance of a place and its worth as a habitat. A good part of people's concern is not about the conservation of natural resources or about cultural significance, as such, but about the natural world as a direct object of value, often quite independent of any use it might have for individuals. This concern has been voiced by philosophers in terms of the 'intrinsic value' of nature and by economists in terms of its 'existence value'. Whether either term has done much to clarify the issues is a moot point to which we return later. What it does signal is the direct response of many to the needless destruction of the non-human environment.

Addressing value conflicts

Value conflicts

Distinct dimensions of environmental good and bad can clearly coexist. Acidification might have effects on forests regarded as a source of timber (economic), as a habitat (biological), and as a socially significant landscape, an object of aesthetic appreciation and source of recreation. On all three dimensions of value the reversal of acidification might count as a good. However, the different dimensions of damage point also to the conflicts that can exist between different kinds of value that might be attributed to the environment. The drainage of marshland from the economic perspective of agricultural productivity and the possibility of increasing sustainable agricultural yields over time might count as improvement; but from the perspective of biodiversity or the cultural significance of ancient marshes it may be damaging. Conversely, a farmer might see the decision to flood as damaging and will worry about the growing influence of conservation policy on the future of his livelihood. From the landscape and recreation perspectives, the decision to destroy rhododendron on the hillsides of Wales might be seen as damaging: from the perspective of protecting the local flora and fauna it is an improvement. A windfarm is both a way of decreasing the loss to the resource base of the economy by the use of a renewable resource and potentially a visual eyesore. Within the same dimensions of value, too, there can be conflicts. To increase the amount of deciduous trees in a forest plantation in the UK may constitute an improvement in the diversity of local flora but threaten the red squirrel who fares less well than the grey in mixed woodland. The policy maker is often faced, not with a clear-cut decision between protection and damage, but with the distribution of different kinds of damage and benefit across different dimensions of value. Moreover, there is a conflict between the avoidance of environmental damage, and other social, economic and cultural objectives. These include not only direct conflicts, say, between the economic benefits of a road development and the environmental damage it will cause, but also indirect conflicts in terms of the opportunity costs of environmental projects, that is, the resources employed that could have been employed for other projects, both environmental and non-environmental.

The distribution of goods and harms

Any decision on such environmental conflicts has a distributional dimension. A decision will take place against the background of a distribution of property rights, incomes and power and it will distribute damage, costs and benefits across different social groups. Hence environmental problems raise issues of equity and justice. To preserve tigers, rhinoceros or elephants through the establishment of a nature park will benefit visiting tourists and might even benefit the animals. But it will often adversely affect the livelihoods of those pastoral and agricultural communities living at the margins of survival. The benefits of increases in the production of greenhouse gases may come to first- and third-world elites, but the costs fall on the poor. The siting of roads, power stations and dumps for toxic waste will damage the quality of life for particular communities. Thus, any decision-making procedure has to be assessed in terms of the potential distributional implications it might have. Who is damaged and who gains the benefits? The environment has added a strong temporal dimension to these distributional concerns. Many of the adverse environmental consequences of human activity that are beneficial to favoured members of our current generation – for example, the use of non-renewable resources – will fall upon future generations. This raises questions about our responsibility for future generations and the inequities many current decisions might have for them.

Addressing conflicts

Environments are sites of conflict between different values and different social groups. They are also sites of conflict within social groups and even within individuals, where they appear as dilemmas. These conflicts occur at a number of different levels – at the local level in the management of environmentally significant sites, at the level of decisions about specific economic and environmental projects, at the level of policy and at the level of regulation. They are conflicts that concern both citizen and policy maker. How are such conflicts to be resolved?

One response to the problem of value conflict is to find a common measure of values through which the gains and losses in different values can be traded off one with another. This position is associated with utilitarianism which, through cost-benefit analysis and welfare economics, has tended to dominate much public policy making. Classically the utilitarian argues that we should aim at the decision that maximises welfare. Hence we need a measure of welfare such that gains and losses in welfare can be appraised and the choice that produces the greatest total welfare be discerned. Thus utilitarianism, understood as an account of decision making, recommends the policy that maximises the welfare of affected agents.

In its modern form, the welfare of agents is often taken to consist in the satisfaction of their preferences, the stronger the preferences the greater the welfare improvement. One putative advantage of this definition from the perspective of modern welfare economics is that it brings welfare improvements and losses under the 'measuring rod' of money: a person's willingness to pay at the margin for an object or state of affairs, or their willingness to accept payment for its loss, provides a measure of the strength of their preferences for it. The different values that conflict are understood within this perspective to be expressions of different preferences which through willingness to pay measures can be bought under a common currency for the purposes of comparisons of different options. The aim of public policy is realised through the use of cost-benefit analysis (CBA) in which welfare benefits are traded off against welfare costs.

This approach has dominated recent environmental policy making, and cost-benefit analysis has been the most widely used decision-making tool in environmental policy making of the last thirty years. It is assumed that individuals in society have preferences whose satisfaction increases their welfare, and that these can be measured by individuals' willingness to pay for their satisfaction. The analyst can thereby simply compute the costs and benefits of any project. The benefits are identified by summing the different amounts that affected individuals are willing to pay for the project, the costs by summing the different amounts affected individuals are willing to pay for the project not to proceed. If benefits outweigh costs then a project is worthwhile. Of a number of projects, the best is that which produces the greatest sum of benefits over costs.

In the first part of this book we examine this dominant utilitarian approach to environmental policy making and find it wanting. In chapter 2 we outline the central assumptions of the approach. Utilitarianism is the doctrine that the best action or policy is that which produces the greatest amount of welfare or well-being of agents. As such the doctrine is welfarist – it holds that the only thing that matters in itself and not as a means to some other end is the welfare of individuals. It is consequentialist – whether an action or policy is right or wrong is determined solely by its results, its consequences. It is an aggregative and maximising approach – we should choose that policy which produces greatest total of welfare. In chapters 2 through to 5 we critically examine each of these assumptions and examine their implications for environmental decision making. In chapter 2, we look in detail at the welfarist assumption. We consider different accounts of the nature of human and non-human well-being. We consider whose well-being might count in consideration of public policy, human and non-human. And we look at whether it is possible to compare changes in the welfare of different individuals. In considering the possibility of such comparisons we will consider the ways in which many welfare economists have attempted to modify utilitarianism to allow for choices without making such interpersonal comparisons of the welfare of different agents. In chapter 3 we consider the consequentialist component of utilitarianism and introduce the reader to the other two central perspectives in ethical theory, the deontological perspective and the virtues-based perspective. We do so by considering two of the central objections that have been raised against consequentialism. The first objection is that it permits too much. There are, it is claimed, some acts, for example acts of torture, that we ought not to do even if it improves general well-being. This claim is often supported by the Kantian proposition that individuals have a moral standing and dignity which rules out certain acts towards them, even if this improves the general welfare. Sometimes this view is expressed in terms of individuals having rights that cannot be overridden for the general welfare. This Kantian perspective has itself been subject to criticism from another direction by communitarian writers who reject the particular form of individualism that it assumes. The second objection is that consequentialism requires too much, that there are acts which may improve the general welfare, but which, since they clash with an agent's deepest commitments, one cannot expect her with integrity to perform. In outlining those objections we thus introduce the reader to the two main ethical perspectives which are offered as alternatives to consequentialism. The first is the deontological perspective which claims that there are constraints on performing certain actions even if they should lead to the most valuable state of affairs. The second is the virtues perspective, which claims that we should start ethical reflection with the question of what sort of person we should be, what excellences of character, virtues, we should develop, and what defects of character, vices, we should avoid. We consider how consequentialists might respond to these objections. We conclude by defending a pluralist perspective in ethics which is developed in more detail in chapter 5. Chapter 4 considers the maximising assumption of utilitarianism, that we should aim to improve total welfare, and discusses some central problems concerning the just and equitable distribution of goods that this position appears to face. It considers in what way our assumptions about equality should be introduced in environmental choices. Chapter 5 examines two distinct assumptions of classical utilitarianism: value monism – the assumption that there is only one thing that is ultimately valuable in itself, and value commensurability – the assumption that there is a single measure of value through which we can arrive at policy choices. In this chapter we criticise both of these assumptions and examine alternative deliberative and expressive accounts of rational choice that are consistent with the recognition of value pluralism and value incommensurability. We consider how the different consequentialist, deontological and virtues-based traditions of ethical theory can take pluralist forms and the different accounts they offer for the resolution of value conflicts. We defend a form of pluralism that rejects a central assumption that recent presentations of those traditions share, the assumption that rational reflection on ethical choices and conflicts requires an ethical equivalent of a scientific theory, complete with theoretical primitive assumptions, from which our specific obligations could be deduced.

The main alternative to the dominant consequentialism in recent environmental ethics centres around the claim that our environmental crisis requires a radically new environmental ethic. This new ethic is taken to require fundamentally new foundational assumptions that break from the anthropocentric assumptions of existing Western traditions of ethical theory. The new theory involves the exten-sion of the class of beings to whom moral consideration is owed and the recognition that non-human nature has intrinsic value. In part two of the book we examine these claims about the need for a new environmental ethic. In chapter 6 we consider the attempt to offer a new ethical theory that extends the domain of moral consideration beyond human beings. In chapter 7 we examine the claim that nature has intrinsic value and the various meta-ethical debates this has raised about the status of ethical claims. We will suggest that the demand for a new environmental ethic shares with its opponents the assumption that we criticised in chapter 5. It assumes that rational ethical reflection requires that we develop a new ethical theory with a few ethical primitives from which our moral obligations can be derived. We suggest that while our environmental crisis might require radical changes to political and economic institutions there is no reason to assume that it requires a new ethic in this sense of a new ethical theory. Such an approach divorces reflection on the environment from the specific ways that environments matter to people, and as such it loses touch both with why it is reasonable to care about the environment and with what is at stake in many environmental disputes. Where it does have an influence it is not always a benign one, issuing, for example, in an over-emphasis on the value of wilderness con-ceived in a particular way. This has been employed in the unjustified exclusion of marginal communities from the places they have inhabited. It has also tended to employ an abstract and thin meta-ethical vocabulary which is blind to the role of place and history in the evaluation of both cultural and natural worlds. In contrast, we argue for the need to begin ethical reflection from the human scale of values evident in our everyday encounters with human and non-human beings and environments with and in which we live. Ethical reflection needs to be embedded in the different kinds of human relation to our environments we have outlined earlier in this chapter, and the thick and rich ethical vocabulary through which we articulate those relations. In chapter 8 we examine what claims, if any, nature has on us in virtue of being natural – the claim that naturalness is itself a source of value. We argue that what does emerge from consideration of 'naturalness' as a value is the role that history and narrative play in our evaluative responses to environments, beings and things around us. However, this role that narrative and history play is by no means confined to the natural, but applies also to our relations to human and cultural landscapes and environments.

In our critical discussions of alternative approaches to environmental values in parts one and two we defend two claims. First we defend a form of pluralism about values which is sceptical of the attempt to understand ethical reflection in terms of moral obligations that are derived from sets of ethical primitives. Second we argue for the importance of history and narrative in environmental valuation. In the third part of the book we develop those claims in more detail and examine their implications for environmental policy making. In chapter 9 we expand on our account of the role of history and narrative in environmental value through consideration of some everyday nature conservation problems. In chapters 10 and 11 we consider the ways in which our approach might be extended to aid in the consideration of issues of policy around biodiversity and sustainability. In chapter 10 we argue that our narrative approach serves as a corrective to the itemising approach to biodiversity policy which is fostered by utilitarian approaches to environmental decision making. In chapter 11 we suggest that our approach offers a basis for understanding the limitations of approaches to sustainability that present the problem in terms of maintaining or improving human and natural capital. Finally in chapter 12 we consider the implications of our approach for an understanding of what makes for good decisions about the environment. We examine recent deliberative responses to standard utilitarian approaches to environmental decision making and argue that while they offer a more plausible basis for understanding what makes for good public decisions, they tend to share with their opponents a picture of decisions as discrete events that are to be appraised as such. We will argue that this picture of decisions is misleading. Decisions are not always discrete events, and even where they are they should not be appraised as such. They can only be properly appraised in terms of historical patterns of choices through which the character of institutions is expressed and developed.

In developing our positive historical approach we do not intend here to give a complete defence of this position. Our outline of an alternative to both the dominant utilitarian approaches in environmental policy making and the opposing position of traditional environmental ethicists aims to be a beginning of a conversation rather than anything close to the last word on it. We further hope for an evolution of this conversation in as spirited a fashion as we have seen with the other two forms of understanding environmental values in the last several decades.

PART ONE

Utilitarian approaches to environmental decision making

2 Human well-being and the natural world

Introduction

In chapter 1 we noted that environmental decisions involve conflicts between different values and interests. The utilitarian tradition approaches the issue of apparent conflicts between different values and interests by attempting to find some measure through which different ends can be traded off with each other so as to maximise the total good. A loss in one area can then be traded against a gain elsewhere. The loss in timber revenues may be shown to be outweighed by a gain in recreation uses of a woodland, or the satisfaction it brings many to know that a bird habitat has been protected. The loss of a habitat through the construction of a road is traded against the gain in travel times. On this approach, given the existence of conflict between different values and interests, one finds some common measure through which losses and gains for different affected agents across different values can be traded off with each other: a decision can be reached as to which option produces the best solution. As we shall see, the utilitarian approach is one that has a large influence in public policy. It provides the theoretical basis of the mainstream economic views of environmental decision making.

The utilitarian approach requires, first, the identification of some common measure of value for determining the significance of an option in relation to other possible options. Second, it requires some criterion for determining what is the 'best' solution given some choice of options. While there are many varieties of utilitarianism that have been developed by philosophers and economists, the original 'classical' articulation of this approach offers the following two answers to these two requirements.

1. Measure of value: The measure we employ should be a measure of the happiness or well-being of those parties affected by a decision.
2. Criterion of best option: The best option is that which, from all available alternatives, has the **consequences** which **maximise** the well-being of affected agents, i.e. the best action is that which produces the greatest improvement in well-being.

If one considers the proposed measure of value and the proposed criterion of what is the best option, it will be evident that the theory makes three distinct claims which characterise this approach and set it off from other accounts of practical choice.

1. It is **welfarist**: The only thing that is good in itself and not just a means to another good is the happiness or well-being of individuals.
2. It is **consequentialist**: Whether an action is right or wrong is determined solely by its consequences.
3. It is an aggregative **maximising** approach: We choose the action that produces the greatest total amount of well-being.

Clearly, however, we need not simply accept this account of value assessment as the right one. Each of these assumptions has been at the heart of several controversies in philosophy, economics and political theory. Each raises distinct lines of questions and possible lines of criticism. If one considers the assumption about welfarism, we might ask the following questions: What is well-being or happiness? How does one measure well-being or happiness? If we can measure it, how can we compare the gains and losses in the well-being of different individuals and groups? Is the well-being of individuals the only thing that is good in itself? And whose well-being counts? How wide do we draw the circle of those whose welfare is relevant in our calculations of relative happiness? If we consider the consequentialist assumption, we might ask the following: Aren't there constraints on actions that are independent of the consequences? Is the rightness or wrongness of an action just a matter of its consequences? Could it ever be right, for example, to torture an innocent child or to punish an innocent woman for the sake of improving the general welfare? Are there actions like torture that no good person could with integrity perform even if it led to a greater total welfare? Finally, if we consider the assumption about maximisation, we might raise the following questions: Should only the total amount of welfare count across different individuals or should there be some way in which each individual's welfare counts in and of itself? What implications would a maximising policy have for the distribution of welfare and considerations of fairness and justice? Doesn't a just distribution of goods matter in and of itself? In the specific case of the environment, would it be right, for example, to decrease the total environmental impact of pollution by concentrating it on already environmentally impoverished groups whose well-being is already lower than that of others?

These questions and others will be addressed over the following chapters in this part of the book. In this chapter we will focus on the first set of questions, those concerning the nature and value of welfare. In chapter 3 we consider arguments around the defensibility of consequentialism. Chapter 4 focuses on the distributional problems with standard utilitarian approaches. Chapter 5 examines the very idea that there is a single measure of value through which different options can be compared. In all we will begin with a general discussion of these assumptions of utilitarianism, and then consider their later elaboration and development for specific environmental choices.

Welfare: hedonism, preferences and objective lists

We start with the welfarist component of the theory. What is it for an individual's life to go well? Classical utilitarianism defended in the eighteenth and nineteenth centuries by theorists such as Jeremy Bentham (1748–1842) and James Mill (1773–1836) is distinguished from later versions of the theory primarily in terms of the hedonistic account of well-being or happiness it gives.

The hedonistic account of well-being

Bentham and the felicific calculus

On the hedonist conception, welfare or happiness consists in pleasure and the absence of pain. The claim that all that matters in deciding social policy is pleasure is given a concise and uncompromising formulation by Jeremy Bentham who is generally recognised as the chief founding author of classical utilitarianism. The account of the principle of utility is found in his *Introduction to the Principles of Morals and Legislation* (1789). The value of pleasure and the disvalue of pain are to be measured by their intensity, duration, certainty and propinquity (or 'nearness'). Units of pleasure accounted for across these dimensions can then enter into a calculus of happiness – the felicific calculus. The best action will be that which tends to produce the greatest sum of pleasure over pain for those affected. Thus consider the following passage from the work:

> To a person considered by *himself*, the value of a pleasure or pain considered *by itself*, will be greater or less, according to the four following circumstances
>
> 1. Its *intensity*.
> 2. Its *duration*.
> 3. Its *certainty* or *uncertainty*.
> 4. Its *propinquity* or *remoteness*.
>
> These are the circumstances which are to be considered in estimating a pleasure or a pain considered each of them by itself . . .
>
> To take an exact account then of the general tendency of any act, by which the interests of a community are affected, proceed as follows. Begin with

any one person of those whose interests seem most immediately to be affected by it . . . Sum up all the values of all the *pleasures* on the one side, and those of all the pains on the other. The balance, if it be on the side of pleasure, will give the *good* tendency of the act upon the whole, with respect to the interests of that *individual* person; if on the side of pain, the *bad* tendency of it upon the whole.

Take an account of the *number* of persons whose interests appear to be concerned; and repeat the above process with respect to each. *Sum up* the numbers expressive of the degrees of *good* tendency, which the act has, with respect to each individual, in regard to whom the tendency of it is *good* upon the whole: do this again with respect to each individual, in regard to whom the tendency of it is *bad* upon the whole. Take the *balance* which if on the side of *pleasure*, will give the general *good tendency* of the act, with respect to the total number or community of individuals concerned; if on the side of pain, the general *evil tendency*, with respect to the same community.

(Bentham 1789: 4. 38–40)

Bentham promises, then, measures of pleasure and pain which make arriving at an individual or public choice into a matter of mathematics. There are a variety of questions one might raise about Bentham's claims in this passage: Are the circumstances listed – intensity, duration, etc. – the only ones that need to be considered in estimating a pleasure or pain of an individual? Do they provide a method for measuring different pleasures for the purposes of arriving at a sum? Is it possible to put different kinds of pleasure and pain on a single felicific scale? How can the pleasures and pains of different individuals be compared with each other? Is it feasible to calculate and evaluate the consequences of all the alternatives whenever we make a moral decision? In posing these questions we are raising some of the major difficulties that confronted the utilitarian. Some of these difficulties were addressed by John Stuart Mill (1806–1873), who defended utilitarianism, modifying it in some respects in the process.

John Stuart Mill

Mill was introduced to Bentham and indoctrinated in his ideas by his father James Mill. At first an ardent disciple, J. S. Mill came to believe that Bentham's conception of human nature and human happiness was much too narrow. But he believed that the modifications that needed to be made were quite consistent with the central idea of utilitarianism. His principal work on this topic, *Utilitarianism* (1861), was a defence of the theory against critics and was intended for the general reader, in the spirit of the reformist and educational ideals of the utilitarians of the nineteenth century. In it Mill endorses the basic utilitarian standpoint:

> The creed which accepts as the foundation of morals, Utility, or the Greatest Happiness Principle, holds that actions are right in proportion as they tend to promote happiness, wrong as they tend to produce the reverse of happiness. By happiness is intended pleasure, and the absence of pain; by unhappiness, pain, and the privation of pleasure . . .
>
> (Mill 1861: 257)

However, Mill rejects Bentham's conception of human happiness by distinguishing between different qualities of pleasure, not just different quantities. It is worth keeping in mind the questions posed above, in order to see what answers are suggested by Mill in the following influential passage:

> It is quite compatible with the principle of utility to recognise the fact, that some kinds of pleasure are more desirable and more valuable than others . . . If I am asked, what I mean by difference of quality in pleasures, or what makes one pleasure more valuable than another, merely as a pleasure, except its being greater in amount, there is but one possible answer. Of two pleasures, if there be one to which all or almost all who have experience of both give a decided preference, irrespective of any feeling of moral obligation to prefer it, that is the more desirable pleasure. If one of the two is, by those who are competently acquainted with both, placed so far above the other that they prefer it, even though knowing it to be attended with a greater amount of discontent, and would not resign it for any quantity of the other pleasure which their nature is capable of, we are justified in ascribing to the preferred enjoyment a superiority in quality, so far outweighing quantity as to render it, in comparison, of small account.

> Now it is an unquestionable fact that those who are equally acquainted with, and equally capable of appreciating and enjoying, both, do give a most marked preference to the manner of existence which employs their higher faculties. Few human creatures would consent to be changed into any of the lower animals, for a promise of the fullest allowance of a beast's pleasures; no intelligent human being would consent to be a fool, no instructed person would be an ignoramus, no person of feeling and conscience would be selfish and base, even though they should be persuaded that the fool, the dunce, or the rascal is better satisfied with his lot than they are with theirs . . . A being of higher faculties requires more to make him happy, is capable probably of more acute suffering, and certainly accessible to it at more points, than one of an inferior type; but in spite of these liabilities, he can never really wish to sink into what he feels to be a lower grade of existence. We may give what explanation we please of this unwillingness; we may attribute it to pride, a name which is given indiscriminately to some of the most and to some of the least

estimable feelings of which mankind are capable: we may refer it to the love of liberty and personal independence, an appeal to which was with the Stoics one of the most effective means for the inculcation of it; to the love of power, or to the love of excitement, both of which do really enter into and contribute to it: but its most appropriate appellation is a sense of dignity, which all human beings possess in one form or other, and in some, though by no means in exact proportion to their higher faculties, and which is so essential a part of the happiness of those in whom it is strong, that nothing which conflicts with it could be, otherwise than momentarily, an object of desire to them. Whoever supposes that this preference takes place at a sacrifice of happiness – that the superior being, in anything like equal circumstances, is not happier than the inferior – confounds the two very different ideas, of happiness, and content. It is indisputable that the being whose capacities of enjoyment are low, has the greatest chance of having them fully satisfied; and a highly endowed being will always feel that any happiness which he can look for, as the world is constituted, is imperfect. But he can learn to bear its imperfections, if they are at all bearable; and they will not make him envy the being who is indeed unconscious of the imperfections, but only because he feels not at all the good which those imperfections qualify. It is better to be a human being dissatisfied than a pig satisfied; better to be Socrates dissatisfied than a fool satisfied. And if the fool, or the pig, are of a different opinion, it is because they only know their own side of the question. The other party to the comparison knows both sides . . . From this verdict of the only competent judges, I apprehend there can be no appeal.

(Mill 1861: 258–260)

Mill's insistence that pleasures differ in quality and not only in quantity is in effect a criticism of Bentham: if he is right there can be no calculus of pleasures of the kind that Bentham assumes. But he suggests an alternative method by which pleasures can be compared: pleasant experiences are desired, and some are desired more than others. Two pleasures as disparate as poetry and the now forgotten game of pushpin may still be compared, and one rated more valuable than the other: the criterion being whether it is preferred or desired more by those who have knowledge of both.

One notable feature of Mill's position is that it opens up the theory of utilitarianism for appreciation of the environment in ways which Bentham's narrow conceptions of pleasure and pain might seem to rule out. Classical Utilitarianism should not be confused with 'utilitarian' in a narrow economic sense: the pleasures of living in an unspoilt natural environment can all count. Consider the following excerpt from Mill's *Principles of Political Economy* which makes apparent the degree to which the utilitarian position is open to an enlarged reading of pleasure and pain which includes environmental concerns:

It is not good for man to be kept perforce at all times in the presence of his species. A world from which solitude is extirpated, is a very poor ideal. Solitude, in the sense of being often alone, is essential to any depth of meditation or of character; and solitude in the presence of natural beauty and grandeur, is the cradle of thoughts and aspirations which are not only good for the individual, but which society could ill do without. Nor is there much satisfaction in contemplating the world with nothing left to the spontaneous activity of nature; with every rood of land brought into cultivation, which is capable of growing food for human beings; every flowery waste or natural pasture ploughed up, all quadrupeds or birds which are not domesticated for man's use exterminated as his rivals for food, every hedgerow or superfluous tree rooted out, and scarcely a place left where a wild shrub or flower could grow without being eradicated as a weed in the name of improved agriculture. If the earth must lose that great portion of its pleasantness which it owes to things that the unlimited increase of wealth and population would extirpate from it, for the mere purpose of enabling it to support a larger, but not a better or a happier population, I sincerely hope, for the sake of posterity, that they will be content to be stationary, long before necessity compels them to it.

(J. S. Mill 1848 IV: 6. §2)

However, there are a number of questions one might ask of Mill's appeal to preferences to distinguish different qualities of pleasure. If the argument is to work, the preferences of an informed person who favours the dissatisfied life of Socrates over the satisfied life of ignorance must rely *only* on comparisons of the quality of pleasure. However, it is far from clear that they do. There are a variety of other considerations for choosing one life over another besides the pleasures they might bring – some of which Mill mentions in the passage in which he introduces the preference-test: pride, the love of liberty and independence, a sense of personal dignity. There are no reasons to assume that these are just a matter of pleasure. A person might prefer to maintain a sense of dignity even where this leads to a loss of overall pleasure, be it a loss in quantity or quality. Mill's test illicitly introduces criteria of goodness other than pleasure, such as personal dignity.

There are also other problems that Mill's position shares with that of Bentham, most notably, the claim that well-being is just a matter of having the right mental states of pleasure and the absence of pain. Classical utilitarianism holds that happiness understood as pleasure and the absence of pain is the only thing that is good in itself. If an action brings about an increase in anyone's happiness that is a good consequence; if it reduces anyone's happiness that is a bad consequence. Happiness is taken to mean pleasure and the absence of pain, which are states of consciousness. So on this view the only things that are good in themselves are states of consciousness. Is this satisfactory? One source of doubt is expressed in the following passage from Robert Nozick:

> Suppose there were an experience machine that would give you any experience you desired. Super neurophysiologists could stimulate your brains so that you would think and feel you were writing a great novel, or making a friend, or reading an interesting book. All the time you would be floating in a tank, with electrodes attached to your brain. Should you plug into this machine for life, preprogramming your life's experiences? . . . Would you plug in? What else can matter to us, other than how our lives feel from the inside? Nor should you refrain because of the few moments of distress between the moment you've decided and the moment you're plugged. What's a few moments of distress compared to a lifetime of bliss (if that's what you choose), and why feel any distress at all if your decision is the best one? What does matter to us in addition to our experiences?
>
> (Nozick 1974: 42–43)

We can imagine a similar experience machine having a fine environmental component that gives us all the experiences of living in some beautiful and biologically rich environment but without the existence of that physical environment itself. Would that be enough? Is the experience of nature what the environmentalist aims to defend? Is the experience as such what most of us want? Surely, it might be argued, we want to actually live in an unspoiled natural environment not just have the experience of doing so. Given the choice, most of us would desire to live in a natural world, not to live in a simulation of it, even if the experiences were identical. The point will be explored in more detail later in the book when we discuss the idea of the value of the 'natural'. There are things that it might be thought are good for us and which we will pursue which are not pleasurable states of consciousness or indeed not conscious states at all. We would rather not be a deceived Truman living on a film set (as in 'The Truman Show') even where the experience of friendship is more subjectively enjoyable than the real thing. We would rather have friends than have the pretence of friends.

How is utilitarianism to be modified to accommodate these facts? One response begins with the observation that in these cases we fail to really get what we want. Our desires remain unfulfilled. We prefer a state of affairs in which our desires are really satisfied to one in which they are not. So maybe we should redefine welfare not in terms of the subjective states we experience, but directly in terms of the satisfaction of our preferences. We can further say that the stronger the preference, the greater the welfare improvement given its satisfaction. So we can restate the Principle of Utility to say that we should maximise the satisfaction of preferences. This is preference utilitarianism, the version of utilitarianism nowadays most generally adopted.

Preference utilitarianism

For the preference utilitarian, then, well-being consists in the satisfaction of preferences, the stronger the preferences the greater the increase in well-being. The best policy will be that which maximises the satisfaction of preferences over their dissatisfaction. Preference utilitarianism has been particularly influential in welfare economics. One major reason for this influence is that the theory allows for the use of *monetary measures* in assessing the value of goods. The basic idea is that we can measure how strong someone's preference is for a good by ascertaining how much they would be willing to pay for its satisfaction at the margin. I reveal my greater preference for the next ice cream over the next bar of chocolate by being prepared to pay more for it.

This idea that we can use money as a measuring rod of how much a good will improve a person's welfare forms the basic assumption of standard environmental economics. Consider, for example, the following expression of this approach:

> The preferences for the environment, which show up as gains in welfare to human beings, need to be measured. It may seem odd to speak of 'measuring preferences' but this is exactly what that branch of environmental economics devoted to **benefit measurement** does. A benefit is any gain in welfare (or satisfaction or 'utility'). A cost is any loss in welfare. We are concerned then with the measurement of the benefits from improvements in, or the costs of reductions in, environmental quality . . . In benefit estimation money is used as a measuring rod, a way of measuring preferences. There are very good reasons for supposing money is a good measure of the gains and losses to people from environmental change. What is important is that money just happens to be a convenient measuring rod.
>
> (Pearce et al. 1989: 52–53)

Many environmental goods are unpriced in actual markets. We do not yet pay directly for goods like clean air or the existence of whales. So how is the money measure to be extended to include free environmental goods? Environmental economists have developed three main approaches:

1. Hedonic pricing methods in which a proxy good in the market, such as property values, is employed to estimate a price for environmental goods. Property prices are in part a function of the quality of the surrounding environment, and by comparing prices of similar properties in different locations one can infer how much persons are willing to pay for a good environment.
2. Time cost methods which employ the costs incurred by individuals to use an environmental amenity to estimate values. One infers the strength of

preferences for, say, a forest or nature park, from how much it costs individuals to travel to it or the income lost in doing so.

3. Contingent valuation methods in which monetary values are estimated by asking individuals how much they would be willing to pay for a good or accept in compensation for its loss in a hypothetical market.

This attempt by economists to use monetary values to capture environmental values is often met by a great deal of resistance amongst non-economists. However, if there are problems with it they are not the obvious ones that are often raised. In particular, like J. S. Mill's utilitarianism, this approach does not rule out environmental concerns. The approach does not limit preferences to only some narrow range of goods that are of immediate use to a consumer. The economist is attempting to measure all the different preferences individuals have. The value of environmental goods for individuals expressed through their preferences is normally divided into *use* and *non-use* values.

Use values refer to the satisfactions that come to individuals from the actual use of a good: for example, a woodland might have use values as a source of timber, or more widely as a source of recreation – as a place for a walk, for a picnic, for seeing birds, and so on. Such use values already include more than the narrowly 'economic' use of environments as a mere resource.

Non-use values include at least three kinds of values: first, what economists call *option values*, those that express preferences individuals have for goods they might use, but do not currently use – I might value a particular woodland because it gives me the option of a particular walk I might want to take even if currently I do not exercise that option; second, *bequest values*, those which express preferences for preserving a good for others, including future generations – I might value a woodland for the recreation it will bring my children or grandchildren; and third, *existence values*, those which express preferences for goods that no-one may actually or potentially use, for example, particular species, habitats, ecosystems and the like – I might have a preference that a marshland containing the red raft spider exist even if I have no personal desire to set foot in the marsh and no particular desire to meet a red raft spider.

The economist is trying then to include a wide variety of preferences under 'the measuring rod of money'. Hence, there is nothing as such necessarily narrowly self-interested about this approach. However, there are other problems with this approach that we will pursue below, and that are illustrated by the following two sets of questions.

The first set of questions concerns the use of money as a way of measuring different preferences. The various value conflicts we outlined in chapter 1 are treated on this approach as conflicts between different preferences which can be traded off with each other through a monetary measure. The consumer preference for hardwood doors can through the measure be set beside the conservationist's preference for the existence of tropical rainforests. The willingness of each to pay for their preferences to be satisfied provides the measure for comparisons. But is this approach adequate? Can money just be treated as a 'measuring rod' of preferences? Can all goods be traded-off with each other? What would be the effect on the distribution of goods and harms if you use individuals' willingness to pay as a measure? What would your response be if you were asked how much you would be willing to pay to save tropical rainforests? What would your response be if you were asked how much you would be willing to accept to give false evidence against a friend in court? Is there, in the end, any sensible measure for trading off different environmental values? We will return to these questions in chapters 4 and 5.

The second set of questions concerns the attempt to define what is good for individuals in terms of preference satisfaction. Why should we assume that satisfying preferences is always to the benefit of the individual concerned? Preference utilitarians tell us to maximise preference satisfaction because they take that to be the best thing for human beings. The main initial problem with the preference-satisfaction theory in this crude form is that it fails to allow the possibility that a person may be mistaken about what is best for her. She prefers one food A, which unbeknownst to her is full of carcinogens, to another food B because she believes, mistakenly, that A is better for her health than B. If her preference is satisfied will she be better off? A person who is seriously ill has a strong desire not to take liquids. Would satisfying that desire improve their welfare? Even if we equate happiness with having one's desires satisfied, is it not possible that you may desire what is not good? In that case it will be possible to be happy while failing to achieve well-being. Indeed, this is what Mill is pointing out in saying that while the fool and the pig may be satisfied with their lot, the life that employs the higher faculties is a better life for a human being. It looks implausible to say that if you satisfy whatever preferences people happen to have then you will make them better off.

More sophisticated versions of preference utilitarianism allow that this is the case. It is not the satisfaction of *any* preferences that improves welfare, but the satisfaction of the preferences of fully informed competent agents. If I had known that tasty bit of food A caused cancer I would not have eaten it. Why? Because I have a settled preference for good health which generally overrides that of taste. My mental competence is disturbed during an illness, and I have no desire for liquids, even though I require them for my health. Had I been fully competent, I would have a preference for liquids. The position allows for error but still holds that whether something is good for a person depends ultimately on what they would want or value. What is good for us is still determined ultimately by our preferences.

Objectivist accounts of welfare

Is this more sophisticated informed preference account of welfare adequate? One line of criticism is that it doesn't adequately deal with the ways in which information can be involved in forming a preference for an object. In some cases information can serve to ascertain whether an object that I currently desire in fact satisfies other given preferences. I have a preference for some food which unbeknownst to me is carcinogenic. Were I fully informed about the food I would no longer prefer it, for I have a settled preference for good health which has priority over my preference for gastronomic pleasure. This role for information is quite compatible with the informed preference account of well-being. However, informing a person can also act in a second way to form or reform her preferences. Education often is not a question of ascertaining whether an object fits current preferences, but rather a matter of altering preferences by pointing to features of the object that make it worthy of being preferred. For example, I may have had no preferences at all for a flat muddy piece of ground by the sea. I then take some walks in the area with a friend who has a great deal of ecological and social knowledge of the place, who points out biological features I had no inkling of, fills me in on its history, and so on. On being educated about salt marshes I may subsequently come to value them a great deal, and this education might make a large difference to my well-being: I walk by the coast with developed capacities to see and hear what is there. But here my well-being is increased, not by allowing me to better realise some given preferences, but rather, through changes in perception and knowledge, by allowing me to form new preferences. That is what education, both formal and informal, is all about.

The problems with the informed-preference account of well-being parallels the difficulties with J. S. Mill's account of why the preferences of an informed or competent agent should be given priority. Mill appeals to a subjective-state account of well-being. One starts from the informed agent because only she is in a position to judge: she knows 'both sides' of the question, whereas the uninformed agent knows only one. That answer is unsatisfactory because it relies on the introduction of some criterion of excellence independent of pleasure itself. The appeal to the quality of pleasures is an illicit way of introducing independent ideals – famously, in the case of Mill, the values of 'human dignity' and of realising our specifically human capacities. The same point is true in the case of the appeal to the informed and competent agent to defend a preference-regarding account of well-being. It serves only to smuggle in criteria that are independent of the preferences themselves. The only plausible reason for starting from the preferences of the 'informed' and 'competent' is one that refers directly to those independent criteria of excellence. For example, if we are considering the value of certain ecological systems, the preferences of the competent and informed ecologist count in virtue of her sensitivity to the objects around her, such that she is better able to make judgements about the value of the different habitats. It is features of the sites she knows about that give us reason to attach greater weight to her pronouncements, not the fact she has this or that preference per se.

On the objectivist view, preferences as such are not what determines welfare. Rather, it is the other way around. We prefer things because we believe they are good. They are not good because we prefer them. To live well is to have or realise particular objective states – particular forms of personal relation, physical health, autonomy, knowledge of the world, aesthetic experience, accomplishment and achievement, sensual pleasures, a well-constituted relation with the non-human world, and so on. What is good for us depends, therefore, on something about us, on what we are like. If we were angels, water and other material conditions of life would not be valuable to us. Neither would friends. But we are not angels. Given the beings we are, such things are valuable. What is of value to us cannot be independent of the kinds of beings we are, and the capacities we have. This is perfectly compatible with a rejection of the preference-satisfaction account, which says that what we desire or value determines what is valuable to us. On an objectivist account we simply can't choose like that. It says, rather, that improving welfare is a matter of realising certain objective states. This is not to say that an individual's desires are for that reason irrelevant. Because autonomy, the capacity to govern one's own life and make one's own choices, is a human good, it may matter that those objective goods be endorsed by a person. One cannot improve an individual's life by supplying resources that are valuable to the individual by some objective criterion, but not in light of the conception of the good life recognised and accepted by that individual: a person's life cannot go better in virtue of features that are not endorsed by the individual as valuable.

One concept that forms part of a more objectivist account of well-being that appears in both everyday and theoretical discussions of welfare is that of needs. Needs claims are of different kinds. Some needs are relative to specific projects. If I am to get to Chicago by tomorrow then I need to take a plane. One might respond to that need claim by asking if I really need to be in Chicago tomorrow. However, some needs claims are not like that. There are some needs that must be satisfied if a person is to have a flourishing life at all, needs whose satisfaction is necessary if the person is to avoid being harmed (Wiggins 1991). For example, basic needs for water, food, shelter, certain social relations and the like are of that kind. One way of capturing the objectivity of needs claims is through consideration of the logic of the concept. The concept of 'need' has different logical properties from that of 'preference'. A sentence of the form 'a needs x' is extensional (i.e. if a needs x, and x is y, then it follows that a needs y); a sentence of the form 'a prefers x to z' is intensional (i.e. it is not the case that if a prefers x to z and x is y that it follows that a prefers y to z). For example, from 'Joseph needs glucose' and 'glucose is $C_6H_{12}O_6$' we can infer 'Joseph needs $C_6H_{12}O_6$'. However, from 'Oedipus prefers to marry Jocasta to any other woman in Thebes' and 'Jocasta is Oedipus's mother', one cannot infer 'Oedipus prefers to marry his mother to any other woman in Thebes'. Whether or not a person needs something depends on the objective condition of the person and the nature of the object, its capacities to contribute to the flourishing of a person. Whether a person prefers one object to another depends rather upon the nature of the person's beliefs about the objects.

An objective list account of well-being does bring with it problems of commensurability. If one allows a plurality of such goods to be constitutive of well-being, this raises the problem of how we are to compare these goods so as to arrive at a decision as to what maximises welfare. What measures for example could there be for comparing the quality of personal relations with accomplishment or with health and autonomy? We return to these problems in chapter 5. In the rest of this chapter we consider two other questions we raised earlier about the welfarist assumptions of utilitarianism, both of which are central to environmental policy. The first is about whose well-being counts. The second concerns how we make comparisons between the well-being of different agents – an issue that has been at the centre of the foundations of cost-benefit analysis, which remains one of the main utilitarian legacies in practical policy making, including environmental policy.

Whose well-being counts?

One of the possible attractions of utilitarianism for environmentalists is its answer to the question of whose well-being counts. For the classical utilitarian, pleasure is good and pain is bad. For the preference utilitarian, preference satisfaction is good. It makes no difference whose pleasure or whose pain. Therefore the utilitarian is committed to impartiality. Anyone who may be affected by an action is to be considered on equal terms with any other. Distance in time or place makes no essential difference. Thus geographically remote people must be considered: partiality to members of one's own nation or ethnic group is ruled out. Temporally remote people – future generations – must be considered: partiality to one's contemporaries is ruled out.

If pleasure is good and pain is bad, or preference satisfaction is good and dissatisfaction is bad, then there is no reason to think that pleasure or preference satisfaction is good only when it is human, or that pain is bad only when it is human pain. Simply to be consistent we must consider the consequences of our actions for all beings that are capable of suffering and enjoyment, not for human beings only. That utilitarianism thus extends moral concern to all sentient creatures has been recognised since Bentham, even if the implications for socially accepted treatment of animals has not been explored until recent times.

The French have already discovered that the blackness of the skin is no reason why a human being should be abandoned without redress to the caprice of a tormentor (see Louis XIV's Code Noir). It may come one day to be recognized, that the number of the legs, the villosity of the skin, or the termination of the *os sacrum*, are reasons equally insufficient for abandoning a sensitive being to the same fate. What else is it that should trace the insuperable line? Is it the faculty of reason, or, perhaps, the faculty of discourse? But a full-grown horse or dog is beyond comparison a more rational, as well as a more conversable animal, than an infant of a day, or a week, or even a month, old. But suppose the case were otherwise, what would it avail? The question is not, Can they *reason*? nor, Can they *talk*? but, Can they *suffer*?

(Bentham 1789: 17. 283)

Generally, utilitarianism extends the constituency of moral concern not just to the well-being of all humankind, but to that of all sentient beings. The philosopher who has done most to bring out these implications of utilitarianism is Peter Singer, whose book *Animal Liberation* (1976) has been particularly influential on the animal liberation movement. His general point, which can be made independently of his utilitarianism, is one that underlies his critique of *speciesism*: the arbitrary assumption that some criterion of moral recognition or evaluation applies only to the interests of human beings, where the basis of that moral evaluation is clearly not unique to humans. Thus in the case of utilitarianism, if pleasure and the absence of pain are what matters in moral evaluation, it would be speciesist to rule that the pleasure and pain of non-human animals is unworthy of moral consideration. Speciesism, Singer argues, is akin to racism and sexism and should be rejected. In turn, pushing this argument further, one can see powerful tools in utilitarianism for opposing practices such as hunting, factory farming and animal experimentation (especially for cosmetics and the like). We discuss the question of the extent of the constituency of moral concern further in chapter 6.

Making comparisons: utilitarianism, economics and efficiency

As we noted in the first chapter, a major practical expression of the utilitarian approach to environmental decision making has been cost-benefit analysis (CBA), one of the most widely used decision-making tools in environmental policy over the past thirty years. CBA normally proceeds by assuming a preference-satisfaction account of welfare. Individuals have preferences whose satisfaction increases their welfare. The strength of preferences for marginal changes to their current range of goods can be measured by individuals' willingness to pay for their satisfaction. For any environmental project the total benefits of the project can be identified by summing the different amounts that affected individuals are willing to pay for the marginal changes the project brings, the costs by summing the different amounts affected individuals are willing to pay for the changes not to proceed. A project is worthwhile if benefits are greater than costs. Given a choice of projects we should choose the project that produces the greatest sum of benefits over costs.

One way of understanding cost-benefit analysis is as an exercise in utilitarianism that aims at maximising the total welfare in society. This is certainly how it is often understood. However, in the theory of modern welfare economics, CBA is not normally interpreted as an exercise in utilitarianism. Rather, economists have understood CBA to issue not in a welfare-maximising outcome in the utilitarian sense, but in an efficient outcome, where efficiency is given a particular interpretation we shall outline shortly. The reason for the shift to the language of efficiency lies in the worry many welfare economists have about the 'scientific' validity of comparing the welfare of different agents – in making 'interpersonal comparisons of utility': Why should there be a problem in making such comparisons? The argument runs along the following lines. For some agent, Joe, with a given budget, Joe's willingness to pay more for a beer than lemonade shows that the beer increases his welfare more than a lemonade. However, nothing can show that for another individual, Janet, with the same budget, who is willing to spend more on a lemonade than Joe is on a beer, the lemonade will produce a greater improvement in welfare for Janet than the beer does for Joe, unless one assumes an 'equal capacity for satisfaction' across the agents. Since that assumption can't be verified it should not form part of economics. Economics should avoid interpersonal comparisons in welfare. Hence the following comment from one of the seminal papers of the economist Lionel Robbins, criticising the use of interpersonal comparisons:

> The assumption of the propositions which did not involve interpersonal comparisons of utility were assumptions which ... were capable of verification [by observation or introspection]. The assumptions involving interpersonal comparison were certainly not of this order. 'I see no means', Jevons had said, 'whereby such comparison can be accomplished. Every mind is inscrutable to every other mind and no common denominator of feeling is possible.'

> (Robbins 1938: 637)

Should we accept this line of reasoning against the possibility of making interpersonal comparisons of welfare? The argument is problematic on two counts. First, it assumes a dubious philosophy of mind expressed in Jevons's passage endorsed by Robbins. Is it true that states of pain and pleasure are inscrutable to others? How would learning the use of the language of pain and pleasure be possible on that account? It is only because we can and do make interpersonal comparisons of pain and pleasure that we can learn to apply the concepts to ourselves. Should we in any case conceive of pleasure as a 'feeling' engendered by an object or action: what is the 'feeling' engendered by reading a good novel? Second, it assumes a particular theory of well-being, one which takes well-being to consist in having the right pleasurable feelings of satisfaction: since such feelings are subjective states, their strength is not open to verification and hence we cannot compare them across persons. While you and I may have identical incomes and I may be willing to spend more on my next beer than you are on your next novel, the pleasure I get from my beer can't be compared with the pleasure you get from your novel. However, this theory of well-being is in error. Given an objective state account of well-being, the problem simply does not arise. The quality of a person's life can be ascertained from how far they are capable of realising those goods constitutive of human flourishing. This is not to deny that there may be problems in making such comparisons – for example, those stemming from the incommensurability of the goods involved and the variability in the lives and needs of individuals. However, putatively inscrutable states of mind are not among them.

We have seen that utilitarianism requires the possibility of comparing the welfare of different agents in order to arrive at a sum of the total amount that will be produced by any given project. Therefore, if interpersonal comparisons of well-being are rejected, the economist needs another way of trading off different costs and benefits without making such comparisons. The most commonly used criterion is that of Pareto optimality. Pareto optimality can be characterised through the idea of a Pareto improvement as follows:

- Pareto improvement: a proposed situation A represents an improvement over another situation B if some one prefers A to B but no one prefers B to A.
- Pareto optimality: a situation A is Pareto optimal if there are no possible Pareto improvements which can be made over it.

The Pareto-optimality criterion is taken to be a criterion of efficiency. A point to notice about the criterion is that one doesn't have to make any comparison between different agents in order to use it. If one is considering whether a new situation is more efficient than an old one, it is enough to know that someone prefers the new situation over the old and no one prefers the old to the new. This feature, however, makes the criterion unhelpful for actual decision making: as is clear in most environmental cases, practical decision making normally involves both winners and losers, not just winners. Welfare economists have therefore introduced a variant of Pareto optimality to solve the problem of making it relevant to actual choices. The variant uses the idea of a potential Pareto improvement and often goes by the name of 'the Kaldor–Hicks compensation test'. John Hicks, who was one of the developers of this idea, introduces it thus:

> How are we to say whether a reorganization of production, which makes
> A better off, but B worse off, marks an improvement in efficiency? . . . [A]

perfectly objective test ... enables us to discriminate between those reorganizations which improve productive efficiency and those which do not. If A is made so much better off by the change that he could compensate B for his loss, and still have something left over, then the reorganization is an unequivocal improvement.

(Hicks 1981: 105)

Accordingly, the test of efficiency to be used in practical decision making is the following compensation test:

- Kaldor–Hicks compensation test: a situation A is an improvement over B if the gains are greater than the losses, so that the gainers could compensate the losers and still be better off.

CBA is then presented as providing a mechanism for showing that a project passes this compensation test. The idea is that it gives one a tool for policy making that does not require comparisons in the levels of utility of different affected parties. We choose that project which produces the highest number of gains over losses.

Is CBA thus characterised defensible? We have already suggested that the arguments against the use of interpersonal comparisons of well-being are not convincing. Another point to note is a feature that CBA shares with utilitarianism more generally. Both CBA and the neoclassical economic theory from which it emerged have their roots in utilitarianism. What they have in common with the general utilitarian position is welfarism – the requirement that there be a single measure of welfare for making comparisons between options. They also share the assumption of consequentialism. On these issues, they stand, and fall, with utilitarianism: if the arguments about consequentialism and commensurability we will raise in chapters 3 and 5 are sound they have force against both utilitarian and standard economic analysis. Where CBA differs from utilitarianism principally is in the criterion of optimality it employs to avoid making interpersonal comparisons of utility. This raises special problems to do with the distributional consequences of CBA both in general and in its assessment of environmental value. Both CBA and utilitarianism have been open to objections in virtue of their failures to deal adequately with the proper distribution of goods. We will examine both in chapter 4.

3 Consequentialism and its critics

Introduction

You will recall from the last chapter that according to classical utilitarianism the right action is that whose **consequences maximise** the **well-being** or happiness of affected agents; in other words, the best action is that which produces the greatest improvement in well-being. As the dominant approach to environmental policy making, in particular through instruments such as CBA, we have in the last chapter discussed the welfarist component of utilitarianism. In this chapter we examine its second component, the commitment to consequentialism. In the next chapter we will turn to the maximising character of the approach.

The central feature of consequentialism is the claim that the value of human actions resides solely in the value that they serve to bring about. In themselves, they have only instrumental value. Whether an action, be it individual or public, is right or wrong depends upon the character of its consequences. What are valuable for their own sake are certain states of affairs that actions might produce. Actions as such, without regard to the states of affairs they bring about, have no ethical value, positive or negative; they are ethically neutral. They can never be wrong in themselves or right in themselves, but are the means to some further value, and ultimately to what is valuable for its own sake, or 'intrinsically good'. A point to note here is that consequentialism as such is quite independent of how what is intrinsically good is characterised. One needn't defend the classical utilitarian view of what is good, the welfare of individuals, in order to be a consequentialist. Nor does one need to defend some maximising account of the good. Consequentialism says that whatever is intrinsically good, the right action is the one that promotes that good.

The central thought that motivates consequentialism is this: What else could determine what is good other than the value it promotes? Wouldn't it be irrational or inconsistent to say – 'I value A, but I don't think I should act to promote that value'? Surely one should act in ways that make the world a better place, given whatever one's conception of a better place is.

To reject consequentialism is to hold that there are acts one ought not to perform even if their consequences are good or even the best. The objection is that there are, or ought to be, constraints on performing certain actions even where they lead to the best outcome. Familiar examples include: killing the innocent, torturing the innocent, lying and promise-breaking. Examples from the environmental sphere might include: depriving human individuals of a decent environment, even if the consequences come out on some scale 'better'; causing pain to humans or non-human animals, say through experiments, even if the suffering of humans and other non-humans is thereby alleviated; destroying a rainforest, or causing the disappearance of the blue whale, even where the consequences are better. Such actions may sometimes have overall better consequences, but it might be argued that this does not make them right.

Why should someone reject consequentialism? What is there about the position that gives grounds for concern? There are at least two central sources of concern: that consequentialism permits too much; and that consequentialism demands too much.

Consequentialism permits too much

> Tell me honestly, I challenge you – answer me: imagine that you are charged with building the edifice of human destiny, the ultimate aim of which is to bring people happiness, to give them peace and contentment at last, but that in order to achieve this it is essential and unavoidable to torture just one little speck of creation, the same little child beating her chest with little fists, and imagine that this edifice has to be erected on her unexpiated tears. Would you agree to be the architect under those conditions? Tell me honestly!
>
> (Dostoevsky 1994: i.2.5.4)

Consider two variations on a familiar example that raises problems for the consequentialist.

Example 1:   A ruthless head of a large corporation operating with the licence of a corrupt political regime is about to order the release of a cocktail of toxic chemicals into a water supply. However, she is devoted to her family. The only effective way to stop her is to kidnap and threaten violence to an innocent member of her family.

Example 2:   A terrorist has planted a bomb in a busy city centre. He refuses to say where it is. However, he is a good family man. The only way to discover where it is planted is to torture an innocent member of his family.

In both cases the action one is stopping will otherwise involve the suffering and death of a large group of people. It might seem, from the consequentialist perspective, that if the calculation of consequences really does show that torturing an innocent individual would prevent more harm than it caused to all those affected, then it is the right thing to do.

To hold that some actions are impermissible in themselves is to hold a deontological ethic.

> **Deontological ethic**: To hold a deontological ethic is to accept that there are constraints on performing certain kinds of actions even where perform-ing those actions brings about consequences of greater value than not performing them. The basic ethical question is 'What acts am I obliged to perform or not perform?'

In general the deontological objection to consequentialism would run that there are constraints on visiting violence and torture on an individual even where it produces greater good. An absolutist in the deontological tradition will hold that some of those constraints rule out certain actions in principle. It is wrong to torture innocent individuals, whatever the consequences. A moderate deonto-logist will hold that those constraints needn't be absolute. If the harm is great enough – the chemicals will cause the deaths of millions, the bomb is a nuclear device – then one might think the constraints are not strong enough to prevent the action. However, the moderate deontologist will want to say that surely there are constraints on actions that operate here which are independent of the total good that the actions produce. Consequences are not the only things that matter.

What is the problem with consequentialism? The moral standing of individuals

What reasons might one offer for such deontological constraints? One answer is that they are grounded on the moral standing of individuals. There are certain things one cannot do to persons even if it leads to a greater overall good. Individuals have a moral standing which cannot be over-ridden for the purposes of promoting the greater good. As John Rawls puts it, 'Each person has an inviolability founded on justice that even the welfare of society as a whole cannot override' (Rawls 1972: 3).

One major source of this line of argument is to be found in the work of Immanuel Kant:

> Now I say that man, and in general every rational being, *exists* as an end in himself, *not merely as a means* for arbitrary use by this or that will: he

must in all his actions, whether they are directed to himself or to other rational beings, always be viewed *at the same time as an end*. All the objects of inclination have only a conditioned value; for if there were not these inclinations and the needs grounded on them, their objects would be valueless. Inclinations themselves, as sources of needs, are so far from having an absolute value to make them desirable for their own sake that it must rather be the universal wish of every rational being to be wholly free from them. Thus the value of all objects that can be *produced* by our action is always conditioned. Beings whose existence depends, not on our will, but on nature, have none the less, if they are non-rational beings, only a relative value as means and are consequently called *things*. Rational beings, on the other hand, are *called* persons because their nature already marks them out as ends in themselves – that is, as something that ought not to be used merely as a means – and consequently imposes to that extent a limit on all arbitrary treatment of them (and is an object of reverence).

(Kant 1956: 2. 90–91)

In this passage Kant argues that individuals cannot be used merely as a means to an end – they are ends in themselves. What does he mean? Kant's basic idea here is that persons have a capacity to make rational choices about their own lives; they have moral autonomy and free will. This fact about persons confers dignity upon them. They command respect. To treat people merely as a means to some other end, be it their own welfare or that of others, is to fail to respect their dignity. In making this point Kant draws a distinction between persons and things. Things include both artefacts that are the results of our will and natural objects which are not persons. Things can be used merely as a means to an end. The way Kant draws the distinction, then, the rest of nature has only instrumental value, whereas human persons who have the capacity for rational choice are ends in themselves.

This Kantian position may offer one significant line of argument against consequentialism. However, it looks an unlikely source of environmental values, in that it clearly restricts what is of intrinsic value to humans – and not even to all humans, but only those who have the capacity for rational reflection and are thus moral agents. However, it is possible to take the general premise of Kant's argument – that there are beings who have moral standing that cannot be over-ridden for the more general good – but to offer a different account of which beings have moral standing. The argument is used, for example, by Tom Regan (1988) to found obligations to all sentient beings. Regan argues that to possess moral standing or inherent value a being must be, not a moral agent, but a subject of a life:

Those who, like Kant, restrict inherent value to moral agents limit this value to those individuals who have those abilities essential for moral agency, in particular the ability to bring impartial reasons to bear on one's decision making. The conception of inherent value involved in the postulate of inherent value is more catholic, applying to individuals (e.g. human moral patients) who lack the abilities necessary for moral agency. If moral agents and moral patients, despite their differences, are viewed as having equal inherent value, then it is not unreasonable to demand that we cite some relevant similarity between them that makes attributing inherent value to them intelligible and non-arbitrary . . . [T]he relevant similarity is what will be termed *the subject-of-a-life criterion*. To be the subject-of-a-life, in the sense in which this expression will be used, involves more than merely being conscious . . . [I]ndividuals are subjects-of-a-life if they have beliefs and desires; perception, memory, and a sense of the future, including their own future; an emotional life together with feelings of pleasure and pain; preference- and welfare-interests; the ability to initiate action in pursuit of their desires and goals; a psychophysical identity over time; and an individual welfare in the sense that their experiential life fares well or ill for them, logically independently of their utility for others and logically independently of their being the object of anyone else's interests. Those who satisfy the subject-of-a-life criterion themselves have a distinctive kind of value – inherent value – and are not to be viewed or treated as mere receptacles.

(Regan 1988: 241 and 243)

Regan's idea here is that any being who has an experiential life of their own cannot be treated as a 'receptacle', as a mere means, in the way that utilitarianism appears to allow. One cannot justify one individual's suffering purely on the grounds that it increases the total good. Beings who are subjects-of-a-life have inherent value and cannot be thus treated. Regan is confident that this covers mammals of more than one year old, and argues that we should give the benefit of the doubt to non-mammalian animals to whom we cannot with certainty deny a degree of consciousness. As for beings who lack any capacity to be aware of whether their life goes well or not (and this would certainly seem to include trees and rivers), Regan does not rule out the possibility of a case being made for their having inherent worth, but is not hopeful. An attempt to extend the range of moral standing more widely to all living things is made by Paul Taylor (1986), and later was extended and amplified by Gary Varner (1998). According to Taylor's 'biocentric individualism', any being capable of pursuing its own good in accordance with its particular nature is claimed to have moral standing, and as such all living things fall within its ambit. The central thrust of his argument is again to extend the respect that in Kantian ethics is owed to persons as rational agents to all living things as 'teleological centres of life'. Just as all persons are ends in themselves that have inherent worth, so also are all living things in virtue of having goods of their own and hence interests. They are owed the attitude of respect that is demanded of all beings that have inherent worth. Whether the attempt to extend moral standing this widely is convincing is an issue to which we will return in chapter 6.

Rights, conflicts and community

The Kantian view that certain individuals are ends in themselves that cannot be treated merely as means to the promotion of the greater total good is often stated in the language of rights. Individuals are said to have rights to certain kinds of treatment which cannot be overridden for other ends. On this view, to assert that an individual has rights is to grant them certain claims, for example, the claim not to be tortured, that are not open to being traded against increases in total human welfare. Where conflicts exist between some harm that an individual can forgo as a right and the general welfare, rights take precedence. In ethical arguments rights are trumps (Dworkin 1977: xi). If one interprets moral rights in this way, then extending rights to include non-humans, be they sentient animals or even other non-sentient biological entities, is to grant them a status such that there are certain acts one cannot subject them to, even if it leads to greater goods overall.

An immediate problem with any generalised appeal to rights is how to adjudicate conflicts where choices have to be made between different rights holders. The point is applied to an environmental case by Pearce and Moran, two economists who defend a utilitarian approach to biodiversity value against an appeal to rights:

> If all biological resources have 'rights' to existence then presumably it is not possible to choose between the extinction of one set of them rather than another. All losses become morally wrong. But biodiversity loss proceeds apace . . . [I]t is essential to choose between different areas of policy intervention – not everything can be saved . . . If not everything can be saved then a *ranking* procedure is required. And such a ranking is not consistent with arguing that everything has a right to exist.
>
> (Pearce and Moran 1995: 32)

It is certainly true that where 'trump values' themselves are in conflict, the language of rights can descend to assertion and counter-assertion. However, the existence of such conflict does not as such entail a utilitarian approach to their adjudication through a trading off of costs and benefits. What is true is that we are owed some account of the nature of the choice under these circumstances. We develop a more detailed account of such choices in chapter 5.

The use of the language of rights to defend the environmental goods of humans or non-humans and to prohibit acts that are damaging has been open to other major criticisms. First there are objections levelled specifically against the attempt to extend the applications of rights from humans to non-humans. Such objections appeal to supposed facts about non-humans that are taken to make it impossible for them to be bearers of rights: for example, that non-humans are unable to *claim* rights on their own behalf. Another set of objections raises issues about the addressee of such rights, and the duties they are taken to involve. If non-human animals have rights, to whom are they addressed and what are the duties that are taken to be correlative of them. Consider the following remark from an early twentieth-century writer:

> Are we not to vindicate the rights of the persecuted prey of the stronger?
> Or is the declaration of the right of every creeping thing to remain a mere
> hypothetical formula to gratify pug-loving sentimentalists?
>
> (Ritchie 1916: 187)

The question is: if non-humans have rights, then doesn't this entail duties on humans to act as ethical policemen in the natural world? The claim that humans do indeed have a duty to act as ethical policemen in the natural world was sometimes voiced in justification for predator-eradication programmes that served narrower human interests in agriculture and game protection. Appeal was made to an Arcadian vision in which, if the lion could not lie down with the lamb, the lamb could at least exist without the lion. The goal was a civilised wild life: 'first the repression of undesirable and injurious wild life; the second, the protection and encouragement of wild life in its desirable and beneficial forms' (J. Cameron cited in Worster 1977: 265–266). Both rights theorists and utilitarians seem in principle to sanction such justifications of predator control. However, few utilitarians or animal rights theorists now defend the view that humans should act as nature's ethical police force. The standard grounds for rejecting this role are that we are not very good at it. The ecological consequences of predator control have thus far produced more suffering caused by the spread of disease and genetic weakness amongst the prey than predators would have inflicted. We currently lack the knowledge and means to do an effective policing job. However, while this may be true, it still appears unsatisfactory as a response. Were it the case that we could do the job well, there still appears to be something objectionable about removing the lion, the wolf and the hawk from the world. One possible source of concern may stem from the belief in a particular form of individualism that both liberal rights theorists and utilitarians share.

This worry is related to longstanding objections to the very use of the concept of rights as a foundation for ethics that have been raised against the individualism of liberal rights discourse and indeed against utilitarianism by conservatives, communitarians and socialists alike. From somewhat different perspectives, these positions share a common scepticism about the particular form of individualism that is taken to be presupposed by liberal rights theory. Many of the criticisms concern the identity of the subjects of rights. Liberal rights theorists from Kant through to Rawls are taken to assume that individuals have an identity that is prior to and independent of their membership of communities. Consider briefly Rawls's account of justice, which we will outline in more detail in the next chapter. In constructing his theory of justice, Rawls attempts to capture the idea of impartiality through the idea of choice under a 'veil of ignorance'. The principles of justice are those that rational individuals would agree upon in a hypothetical situation in which mutually disinterested individuals choose principles behind a 'veil of ignorance' of their social position, characteristics and conception of the good:

> It is assumed that parties do not know certain kinds of particular facts. First of all, no one knows his place in society, his class position or social status; nor does he know his fortune in the distribution of natural assets and abilities, his intelligence and strength, and the like. Nor again does anyone know his conception of the good, the particulars of his rational plan of life, or even the special features of his psychology such as aversion to risk or liability to optimism and pessimism.
>
> (Rawls 1972: 137)

The very possibility and intelligibility of such a choice has been questioned by communitarian writers such as MacIntyre (1986) and Sandel (1982), because of the view of the self it assumes. It is taken to assume a radically disembodied and socially disembedded conception of the self. Against this view the communitarian claims that an individual's identity is constituted by her conception of the good and her ties to others. An individual's identity is partly constituted through their membership of various communities. An outline of some of these objections is presented thus by Ted Benton in discussing Regan's defence of animal rights:

> It is generally true of the liberal tradition that it assigns moral priority to securing the integrity and autonomy of the individual person or 'subject'. This moral priority is expressed in Regan's preparedness to assign inherent value to subjects-of-a-life, and to recognize inherent value as grounding rights as *prima facie* valid moral claims. I do not take this order of moral priorities to be exclusive to the liberal tradition, or as definitive of it. Indeed, assigning moral priority to the well-being of individuals is . . . widely shared by both liberals and their communitarian critics. It is also shared with the socialist critique of liberal rights theory . . . What I do take to be definitive of liberal views of rights is a certain range of answers to two further questions: (1) What is presupposed in the integrity or autonomy of the individual subject? and (2) What is the relationship between the

moral discourse of rights and justice, on the one hand, and the protection of the integrity and autonomy of the individual subject on the other. Both contemporary communitarian and 'traditional' socialist critics of liberal rights find liberalism wanting primarily because of its distinctive and, in their view, unsatisfactory ways of answering these two questions . . . Whereas . . . Marx's critique focused upon the imputed self-interestedness of the subjects of liberal rights theory, the contemporary communitarian view goes deeper. It calls into question the possibility of the individuation of the self, independently of, or prior to, community membership.

(Benton 1993: 102–103)

These critical perspectives on liberal rights theory are often taken to have special significance in the environmental context. They have implications both for the claim of human rights to environmental goods and for the extension of rights to non-humans. With respect to the former it might be argued that the identity of individuals has not merely a social but also an environmental dimension: an environment is not just something of instrumental value or a physical precondition of human life. Rather, an individual's identity, their sense of who they are, is partly constituted by their sense of belonging to particular places. Particular places, whether 'natural' woodlands, streams and ponds, or 'urban' city streets, parks and quarries, matter to individuals because they embody the history of their lives and those of the communities to which they belong. Their disappearance involves a sense of loss of something integral to their lives. The argument then runs that if one treats individuals as having an identity that is prior to and independent of such attachments one will not be properly able to capture this dimension of environmental concern. Whether this constitutes a reason for rejecting rights as such, or only a particular conception of rights, remains a moot point. With respect to the extension of rights to non-humans it might be similarly argued that it is individuals as members of communities – including ecological communities – that matter, not just individuals per se. We will return to these concerns in the second part of this book.

Consequentialism demands too much

So far, our criticism of consequentialism has focused on the claim that it permits too much. There are actions we ought not to do even if they have the best consequences. We have examined one possible source of this problem, that individuals have a moral standing such that there are certain things one cannot do to persons even if it leads to a greater overall good. We have seen that one articulation of this view is that individuals have rights, and we have considered some objections to liberal theories of rights. In this section we turn to a second criticism that is levelled against consequentialism – the criticism that consequentialism demands too much. Here is an example that Bernard Williams (1973: 97–98) uses to illustrate the point:

> George is an unemployed chemist of poor health, with a family who are suffering in virtue of his being unemployed. An older chemist, knowing of the situation, tells George he can swing him a decently paid job in a laboratory doing research into biological and chemical warfare. George is deeply opposed to biological and chemical warfare, but the older chemist points out that if George does not take the job then another chemist who is a real zealot for such research will get the job, and push the research along much faster than would the reluctant George. Should George take the job?

For the consequentialist, given any plausible account of the good, the right thing to do is obvious: George should take the job. That will produce better consequences both for his family and the world in general. However, to take the job would appear to undermine George's integrity. He must treat his own projects and commitments as just so many desires to be put into the calculus with others.

> It is absurd to demand of such a man, when the sums come in from the utility network which the projects of others have in part determined, that he should just step aside from his own project and decision and acknowledge the decision which the utilitarian calculation requires. It is to alienate him in a real sense from his actions and the source of his action in his own convictions. It is to make him into a channel between the input of everyone's projects, including his own, and an output of optimific decision; but this is to neglect the extent to which *his* actions and *his* decisions have to be seen as the actions and decisions which flow from the projects and attitudes with which he is most closely identified. It is thus, in the most literal sense, an attack on his integrity.
>
> (Williams 1973: 116–117)

What is the problem with consequentialism? Agent-based restrictions on action

How might the consequentialist respond to Williams's objection? Surely, like the older chemist, the consequentialist might tempt the agent in terms of his or her own commitments: 'Look if you are really opposed to chemical weapons, you want to do all you can to stop their development, and that's best achieved by your taking the job. That is what it is to be committed to opposing them.' What can George say? If he is to retain his integrity he has to resist the consequentialist temptation. He has to say something like: 'Even if that is true, I don't want to be the kind of person that could do that. Regardless of the consequences, I won't collude with something to which I am opposed. I refuse to engage in making chemical weapons. There are some things I simply won't do.'

This line of argument against consequentialism focuses on the constraints on what we can expect of an agent performing an act. The argument would run that while it may be 'better from an impartial perspective' that there are fewer chemical weapons or, to use our earlier examples, that one innocent person suffers from torture rather than several innocent people suffer the consequences of poisoned water or an explosion, there are constraints on what we can oblige an agent to do to realise those goals. No one can oblige a person to do acts that run against their deepest commitments, be this research on chemical weapons or torturing an innocent individual.

On what grounds can a person make that kind of claim? One justification for the claim would involve appealing to a 'virtues ethic'. There are acts I cannot do as such, because I do not want to be the kind of person that can do them. Virtue ethics is defended in classical philosophy by Plato and Aristotle (Aristotle 1985) and has undergone something of a recent revival (Anscombe 1958; Crisp and Slote 1997; MacIntyre 1986). On the standard presentation, to hold a virtues ethic is to take the question 'what kind of person should I be?' to be the basic question in ethical deliberation. Primitives of ethical appraisal include the excellences of character, the virtues.

> **A virtues ethic**: The basic ethical question is 'what sort of person should I be?' and an answer to that question cites excellences of character – virtues – that are to be developed, and defects of character – vices – that are to be avoided.

On this line of argument, integrity, the virtue of our chemist, is a basic value, which cannot be overridden by consequentialist considerations, since it is one of the basic excellences of character. Integrity is a central virtue, because it is a condition of having other virtues. It is closely related to the Socratic concept of courage as the virtue concerned with having a sense of what is important and staying firm to it. However, like courage, it is only an excellence of character when in the company of other virtues. Consider the same story told by Williams re-written from the perspective of the zealot for chemical weapons. Would integrity be a virtue? Integrity might be the last quality one would hope the zealot possessed. The sooner he is tempted to betray his beliefs for a comfortable job in advertising the better.

Virtues and environmental concern

A virtues ethic offers another possible line of defence of environmental concern (Barry 1999; O'Neill 1993; Hursthouse 2000; Cafaro and Sandler 2005; Cafaro

2006). On the virtues view we begin with the question of what it is for us to do well as human agents, what sorts of person should we be? In answer, we specify a certain range of dispositions of character that are constitutive of a good human life: sensitivity, courage, loyalty, good judgement, and so on. Many of those dispositions of character will be exhibited in having proper responses to other humans. For example, someone who perceives the undeserved pain of a fellow human being and feels no response of sympathy or compassion will lack one of the dispositions of character that make for a good human life. In an extreme case they would exhibit a psychopathic character. Someone with those dispositions would miss out on many of the goods of human life such as friendship, love, relations with kin and human solidarity. However, there is no reason to assume that such dispositions of character that are part of what makes for a good human life should involve only dispositions to respond to other human beings. The good human life is one that includes the dispositions and capacities to respond appropriately to beings in the non-human world. This will include sensitivity and compassion towards the suffering of other sentient beings. However, it might also involve a wider set of dispositions and responses to the non-human world, attitudes of awe towards the larger universe of which we are just a part, or of wonder at the complexity and interdependence of particular places and living things.

Consider the following passage from the classical defender of a virtues ethic, Aristotle:

> [I]n all natural things there is something wonderful. And just as Heraclitus is said to have spoken to his visitors, who were waiting to meet him but stopped as they were approaching when they saw him warming himself at the oven – he kept telling them to come in and not worry, for there are gods here too – so we should approach the inquiry about each animal without aversion, knowing that in all of them there is something natural and beautiful.
>
> (Aristotle 1972: bk 1, ch. 5)

On this view part of the value of knowledge lies in its development of our capacity to contemplate that which is wonderful and beautiful. Such contemplation extends our own well-being since it realises our characteristic human capacities. There is a relationship between our capacity to appreciate the value of the natural world and human well-being. This Aristotelian position is developed further by Marx in his remarks on the 'humanisation of the senses' in the *Economic and Philosophical Manuscripts*. Both art and science humanise the senses in that they allow humans to respond to the qualities that objects possess. We respond in a disinterested fashion – and it is a characteristic feature of humans that they can thus respond to objects. In contrast, our senses are dehumanised when we respond to objects only as items that satisfy narrowly conceived interests:

Sense which is a prisoner of crude practical need has only a restricted sense. For a man who is starving the human form of food does not exist, only its abstract form exists; it could just as well be present in its crudest form, and it would be hard to say how this way of eating differs from that of animals. The man who is burdened with worries and needs has no sense for the finest of plays; the dealer in minerals sees only the commercial value, and not the beauty and peculiar nature of the minerals; he lacks a mineralogical sense; thus the objectification of the human essence, in a theoretical as well as a practical respect, is necessary both in order to make man's senses human and to create an appropriate human sense for the whole of the wealth of humanity and nature.

(Marx 1844: 353–354)

Those who can respond to objects only in terms of how far they impinge on narrowly commercial interests, fail to develop their specifically human capacities of perception. The developer sees not a wood or forest, but an obstacle to a highway. She sees not a landscape or a habitat, but space for buildings. Persons driven by narrowly commercial interests respond not to the 'beauty and peculiar nature' of objects, to 'the whole wealth of . . . nature' but to the world as an object for the satisfaction of a narrow range of interests. They exhibit not virtues but vices of character.

A response to the objects of the non-human world for their own qualities forms part of a life in which human capacities are developed. It is a component of human well-being. It is in these terms that the specific virtues produced by certain forms of theoretical and practical education can be understood. Education involves not simply the apprehension of a set of facts, but also the development of particular intellectual skills and virtues, and capacities of perception. The trained ecologist, be she amateur or professional, is able to see, hear and even smell in a way that a person who lacks such training cannot. The senses are opened to the objects around them. By starting with the question of what kind of life is a good life to lead, one opens up the room for a wide set of responses to the non-human world.

Consequentialist responses

In this chapter so far we have outlined two sets of objections to consequentialism. The first is that consequentialism permits too much: there are constraints on performing certain kinds of actions even where those actions produce a greater value than not performing them. We considered the view that such constraints are grounded on the moral standing of individuals, that there are things one cannot do to individuals even if it leads to a greater overall good.

Individuals have rights that cannot be overridden by the general good. The second objection is that consequentialism demands too much. There are actions we cannot oblige agents of integrity to perform. We considered two different perspectives on ethics that might emerge from these lines of argument – a deontological ethic and a virtues ethic and considered their implications for environmental matters.

What responses might the consequentialist make to the two lines of argument that we have outlined? The basic consequentialist response is to ask what justification might be given for respecting individuals' rights or integrity. And the consequentialist will further insist that the answer has to be given in consequentialist terms: the world in which rights and integrity are respected is a better world than one in which they are not. There are, in other words, good consequentialist reasons for respecting rights and integrity. There are two possible lines of argument for this position.

Indirect utilitarianism

The starting point for indirect utilitarianism is the observation that there are a number of aims in life which you will only realise if you do not actively pursue them. This may sound paradoxical, but the paradox is a version of a familiar one in everyday life. Consider the paradox of success, for example, the case of individuals engaged in a sport who are so desperate to win that it interferes with their game. The advice one might offer is that if you want to win, don't make winning your aim, just enjoy playing the game. Or again consider the paradox of the hedonist, the person who in making pleasure their aim, misses out on pleasure (think of the partygoer, who is so desperate to have a good time, that they never have one). Again the advice would be that if you want enjoyment, don't go around trying hard to enjoy yourself. Or finally, consider the paradox of self-realisation, of individuals who in aiming 'to realise themselves' end up as self-absorbed bores with nothing to realise. Individuals who realise themselves are normally those whose aim is the pursuit of some project for its own sake, such as a political commitment, intellectual or artistic activity, the good of a community. The basic thought behind indirect utilitarianism is that the same paradoxical pattern holds for the realisation of the greatest welfare: if you make that your aim in particular decisions you won't realise that good. Hence, there are good utilitarian reasons for not using utilitarian calculation in making particular decisions. Indirect utilitarianism doesn't rule out the possibility of utilitarian reflection altogether. One will, however, use utilitarian calculation only in reflecting in a cool hour on what rules and habits it is best to adopt for general use.

Indirect utilitarianism plays on the distinction between utilitarianism as a criterion of rightness and utilitarianism understood as a decision procedure. That is, the indirect utilitarian distinguishes between (a) the criterion by which an action or policy is judged to be right and (b) the decision-making method that is used, on a particular occasion, to decide which action or policy to adopt. The indirect utilitarian says that the criterion (a) is the Principle of Utility, but that the best way to ensure that an action satisfies this criterion may be to apply a simple principle, or to follow one's moral habits. The distinction matters since it means that utilitarian outcomes are unlikely to be achieved by following 'utilitarian' decision-making procedures.

Why should that be? One part of the answer lies in the constraints on information and time that are invariably involved, and the ensuing uncertainty: it will be impossible to calculate and weigh all the consequences of all the alternatives. Hence, the argument goes, it is better to employ some basic principles to guide action. A second part of the answer appeals to the indirect results of the absence of rules of justice, for example, the insecurity this entails. Third there exist collective choice problems. In a number of situations, if individuals pursue actions that aim to maximise happiness, the result will not maximise happiness. Consider, for example, paradoxes of altruism. In a marginally overcrowded and sinking life-boat each individual thinks they will make the difference and jumps out to maximise the total good. The result is that everybody drowns. Some standard rule of priority, say of the youngest to stay, will produce a better result than each acting in terms of a utilitarian calculation. Finally there are the effects on the agents, that engaging in consequentialist reasoning may corrupt the individuals' moral sensibilities and dispositions. The best utilitarian outcomes are produced by communities that are not inhabited by calculative instrumental reasoners. Hence, it may be that there are good utilitarian reasons not to use utilitarian decision-making procedures.

For example, the indirect utilitarian might argue that the best decision rule for achieving the maximum of welfare accords rights that recognise the special importance of certain basic interests as essential components of happiness. Consider the following argument of Mill for rights to protect those interests of an individual which are fundamental to his or her welfare.

> When we call anything a person's right, we mean that he has a valid claim on society to protect him in the possession of it, either by the force of law, or by that of education and opinion. If he has what we consider a sufficient claim, on whatever account, to have something guaranteed to him by society, we say he has a right to it . . . Justice is a name for certain classes of moral rules, which concern the essentials of human well-being more nearly, and are therefore of more absolute obligation, than any other rules

for the guidance of life; and the notion that we have found to be of the essence of the idea of justice, that of a right residing in an individual, implies and testifies to this more binding obligation.

(Mill 1861: 5. 309 and 315–316)

A point to note here is that indirect utilitarian arguments can apply to public as well as private decisions. If the indirect utilitarian is correct, there might be good utilitarian reasons not to use decision-method procedures like cost-benefit analysis in coming to environmental choices, but rather to follow basic principles of intra-generational and inter-generational equity, principles of respect for non-humans and principles which express the sentiments of a community about the places they inhabit. If indirect utilitarianism is correct, then it provides utilitarian arguments against many of the standard utilitarian decision procedures employed in public life.

Extend the account of the good

A second consequentialist response to the objections raised is to argue that the standard counter-examples we have cited point to problems with other elements in the utilitarian theory, not to consequentialism. The problem raised by our counter-examples – such as torturing the innocent or corrupting an individual's integrity – may lie in the particular account of what is intrinsically valuable, viz. the well-being of individuals, or with the idea that we should simply maximise whatever is good, rather than – say – distribute it equitably. Such counter-examples may raise problems for the welfarist and maximising components of utilitarianism, not the consequentialism. Thus one might expand the account of what is intrinsically valuable and say that respecting certain rights, excellences of human character, equality, and so on, are valuable in themselves. The counter-examples show up a restricted account of what the ultimate goods are, not a problem in consequentialism as such.

How far does this response go towards avoiding the standard counter-examples raised against utilitarianism? As we noted above, one of the main thoughts behind consequentialism is how it could possibly be better not to act in a way that makes the world a better rather than a worse place. Imagine then that we change our account of the ultimate goods and take 'a better world' to mean 'a deontologically better world' in which fewer impermissible acts are performed – fewer acts of torture, injustice, etc. – or 'a virtuously better world' – a world inhabited by better, more virtuous people in better relations with each other. However, the question would still arise as to whether and under what circumstances one could with integrity commit an act of injustice to stop wider injustice, or commit a vicious act to prevent further viciousness. A deontological or virtues-based ethic might still want to place constraints on those acts.

Ethical pluralism and the limits of theory

In this chapter we have assessed and criticised the consequentialist component of utilitarianism. In doing so we have outlined two alternative perspectives on ethical choices, deontology and virtues ethics. As they are normally described, the consequentialist, deontological and virtues-based perspectives express incompatible moral theories of a certain kind. They are moral theories which are reductionist in that they offer different accounts of the primitive concepts of ethics and they attempt to show how other ethical concepts can be either defined or justified in terms of those primitive concepts. The primitives can be expressed in terms of the question each takes to be basic in ethics, and the answer each provides.

> **Consequentialism**: 'What state of affairs ought I to bring about?' What is intrinsically good or bad is a state of affairs: actions and states of character are instrumentally valuable as a means to producing the best state of affairs.

> **Deontology**: 'What acts am I obliged to perform or not perform?' What is intrinsically good or bad are certain acts we are obliged to perform: states of character are instrumentally valuable as dispositions to perform right acts.

> **Virtues ethic**: 'What kind of person should I be?' The basic good of ethical life is the development of a certain character. A right action is the act a virtuous agent would perform; the best state of affairs is one that a good agent would aim to bring about.

Thus characterised, these are incompatible positions. However, the incompatibility is the result of their sharing a reductionist account of ethical theory – the view that there is a single basic question and answer in ethical deliberation about the priority of states of affairs, acts and agency. The background assumption is that philosophy should offer us an ethical theory which has ethical primitives that we can apply to find an unequivocal answer to practical problems. A final response to the debate is to reject the assumption that philosophy can and should offer ethical theories of this reductionist kind, each with its competing ethical primitives. (Indeed, some classical theories that come under the name of virtue ethics involved no such reductive assumptions (Annas 1993).) To deny that such theories are either possible or desirable is not to deny that there is a role for either rigorous philosophical reflection about ethics or indeed for some systematising of our ethical beliefs into what might be called theories. It is to deny that there exist primitives in ethics to which all others must be reduced. There exist a plurality of basic concepts and perspectives in ethical deliberation which cannot be reduced to each other. If one takes this view, then one must allow that these can come into conflict with each other, and where they do there exist tragic choices and loss. For example, hitherto virtuous people can find themselves faced with doing terrible acts to save the good, while admirable people act with integrity in contexts where to do so is hopeless and leads to no good. Tragedy in human life may simply be ineliminable. We return to consider this claim in chapter 5.

4 Equality, justice and environment

As has already been pointed out in earlier chapters, according to utilitarianism the right action is that whose consequences maximise the well-being or happiness of affected agents. In the last two chapters we have looked at the welfarist and consequentialist components of the doctrine. In this chapter we are concerned with the aggregative maximising component – the claim that we should choose the action that produces the greatest total amount of well-being. Should maximising well-being be all that matters? What implications would this maximising approach have for the distribution of welfare and considerations of fairness and justice?

In the context of environmental decisions, distribution clearly does matter. It matters both to the processes that lead to decisions and to the consequences of decisions. Decisions are often shaped in the context of an uneven distribution of wealth, power and voice. The consequences of decisions are often distributed unevenly across class, gender and ethnicity. Environmental harms often fall most heavily on the poor (Martinez-Alier 2002). The siting of toxic-waste dumps and incinerators, of open-cast mines, roads, runways and power stations will all have adverse effects on the communities that are forced to live with them. The construction of dams can lead to the impoverishment and social dislocation of communities displaced by flooding. The introduction of large commercial fishing fleets will lead to the loss of fishing stocks for local and small-scale fishermen operating in traditional fleets. Nor is it only environmentally damaging policies that have such distributional consequences. So also can policies that purport to be environmentally friendly. Forest regeneration projects that prohibit the cutting of trees for firewood may create more woodland, but they can increase the burden on women who are forced to travel further to collect wood, or who suffer disproportionately from the health impacts of alternative fuels (Agarwal 2001). Nature conservation programmes that involve the creation of national parks can lead to the impoverishment and dislocation of communities who are excluded from the parks, or whose pastoral and hunting activities are severely curtailed (Guha 1997). Biodiversity policies can also raise significant distributional issues. The introduction and commercial exploitation of intellectual property rights over genetic resources can benefit large commercial organisations at the expense of those communities in which much of the genetic diversity is to be found. The uneven distribution of power and of harms also has a temporal dimension. Many adverse consequences of current economic activity will be felt by future generations: the possible effects of the depletion of non-renewable resources, the risks from toxic and radioactive materials, the disappearance of sources of wonder in the non-human natural world and of places that offer continuity with the past. Future generations, of necessity, lack power and voice. So also do non-humans. Many of the results of environmental damage are visited upon the non-human world: the disappearance of habitats, the pollution of waters, the introduction of intensive industrial patterns of farming as well as the continuation of agricultural practices that involve intense suffering for domestic animals.

Apparent from these examples is the scope of the distributional dimensions that environmental problems raise. The environmental consequences of different activities are distributed unevenly across class, ethnicity and gender, within and across national boundaries, within and between generations, and between species. Even for those forms of environmental change, like global warming, that are 'global' in the sense that they potentially affect all beings, current generations, future generations and non-humans, the nature of the effects will rarely be uniform. The effects can be mitigated in the richer parts of the globe in ways they cannot in the poorer; the effects of flooding will be more severe on coastal communities than on inland communities; adverse effects on current generations are likely to be less severe than those on future generations. A central and unavoidable dimension of environmental value, then, concerns the distribution of power and property between different groups and the subsequent distribution of goods within generations, across generations, and across species. In this chapter we consider how well the utilitarian and economic approaches outlined in the last two chapters deal with these distributional dimensions.

Utilitarianism and distribution

For the utilitarian the distribution of goods or welfare has only instrumental value. We should choose that distribution of goods that maximises the total amount of well-being. That feature of the theory appears to allow some deeply inegalitarian implications. Consider, for example, the following justification offered by one of the policy makers involved in the decision to build the Narmada Dam in India which left large numbers of peoples displaced from their homes:

Then there was the question of the oustees – the trauma of the oustees. Now this was also given. Earlier we had not mentioned that there is indirect loss, but mind there is indirect benefit also. As I pointed out in the report, benefit is in the sense that if a person is uprooted from a place he suffers a trauma because of displacement, but at the same time, there are other people who gain because the water comes to them ... In fact the beneficiaries are more than the sufferers. Therefore the quantum should at least be equal ...

(quoted in Alvares and Billorey 1988: 88)

The argument here is a classical utilitarian argument. There is the suffering and social dislocation of those uprooted from the place in which they have lived. But at the same time, there are other people who gain because the water comes to them. Since the beneficiaries are far more numerous than those who suffer, the total welfare gains should at least be equal to the suffering, and in fact will be more.

Graphically the argument can be represented as follows:

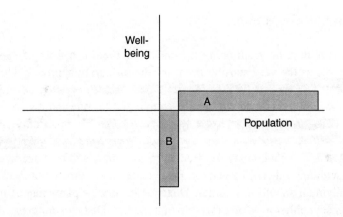

Figure 4.1 Distributions of Welfare and a Utilitarian Justification of the Narmada Dam

As long as the area of the block A – the benefit to that population – is greater than the area of block B – the suffering caused to the oustees – the project is justifiable.

A number of questions might be raised about this sort of utilitarian reasoning:

1. Do the sums actually come out as claimed? For example, are the benefits of the project exaggerated, and the costs underestimated?
2. If the sums did come out right, would it be justifiable to knowingly cause intense suffering to a minority in order to increase the welfare of a larger population? Is the total amount of welfare all that matters in itself?
3. Can you do sums like this at all in the first place? Is there a measure of value that allows you to do moral mathematics like this?

The first, empirical, question is an important one, and there is certainly strong empirical evidence for the claim that in cost-benefit analysis for large projects costs are underestimated (Flyvbjerg et al. 2002; for an excellent sceptical account of the standard calculations for dams see McCully 1996). However, this first question will not concern us here. Our concern in this chapter is with the second question – about the defensibility of the maximising component of utilitarianism. In the next chapter we will consider the third question.

Initially what looks problematic about the utilitarian approach is that it is insufficiently sensitive to the distribution of benefits and suffering. Indeed, utilitarian approaches appear to be incompatible with the requirements of distributional equity and justice. Two central points might be made in response to the claim that utilitarianism is inegalitarian.

Equality in moral standing

The first is that, far from being incompatible, there is a form of egalitarianism that serves as the very starting point for the utilitarian approach. Utilitarianism assumes each sentient individual has *equal moral standing*, as in Bentham's slogan for utilitarian calculations: 'each to count for one and none for more than one'. Thus the utilitarian is committed to a form of impartiality. All sentient beings matter equally in that we should give equal consideration to their interests. Anyone who may be affected by an action is to be considered on equal terms with any other. The gender, race, social class, time of birth or species of the individual should not matter. Distance in time or place makes no essential difference; neither does membership of a species. Distant strangers in other parts of the world, distant future generations, all count if they are affected by an action. So also do some non-human animals. If pleasure is good and pain is bad, or if preference satisfaction is good and dissatisfaction is bad, then there is no reason to think that pleasure or preference satisfaction is good only when it is human pleasure and satisfaction, or that pain and dissatisfaction is bad only when it is human pain and dissatisfaction. Simply to be consistent we must consider the consequences of our actions for all beings that are capable of suffering and enjoyment, not for human beings only.

However, equality in *standing* in this sense does not as such entail equality in the *distribution* of goods and harms. The claim that each sentient being counts equally does not entail the claim that welfare, well-being or resources should be distributed equally. For the utilitarian, we give equal consideration to all interests in the sense that when we consider the total utility, each interest is given equal weight. Each being's 'suffering be counted equally with like suffering . . . of any other being' (Singer 1986: 222). However, this means that where the total satisfaction of interests is maximised by allowing a few to suffer then there is no reason to give, and indeed there is reason not to give, special consideration to the interests of the few. Nothing in the concept of equality of standing, as that concept is understood in the utilitarian tradition, rules out the kind of reasoning illustrated in the Narmada Dam.

Indirect utilitarian arguments for distributive equality

The second line of response is an indirect argument – that, other things being equal, an equality in distribution will tend to produce the greatest total welfare. The indirect argument from utilitarianism to egalitarianism in the distribution of some goods appeals to what is known as their 'decreasing marginal utility'. The marginal utility of a good is the amount of happiness or well-being one gets from the next small increase in it. The idea of decreasing marginal utility is basically that the more you get of some good the less improvement you'll get from the next bit. The third apple you can have offers a lower prospect of improving your well-being than the second, and the second lower than the first. Likewise a person with £1000 is likely to get less welfare improvement from an additional £1 than a person who has but £5. The richer a person is, the less additional utility he or she will get from a unit of income. Therefore, other things being equal, a re-distribution of income from the rich to the poor will increase the total happiness. An egalitarian distribution of incomes will tend to increase total utility.

Figure 4.2 below illustrates the utilitarian argument for equality from decreasing marginal utility (Sen 1997: 17–18). Assume that we are to distribute income between two individuals A and B. The total income to be divided between them is AB. A's share of the income is measured in the direction right from A, B's share is measured in the direction left from B. The marginal utility schedule of A is measured by aa', the marginal utility schedule of B by bb'. If we assume that A and B get the same amount of good or welfare from each unit of income, then the two lines aa' and bb' will be mirror images. The maximum total welfare is realised if the income is divided equally at C, where AC equals BC.

Does this argument resolve the problems we have raised? It does not, for two reasons. First, it does not as such rule out distributing harms to minorities to improve the total welfare. One can allow the intense suffering that the poor and marginalised population moved by the dam will experience and the sums will still come out as indicated in Figure 4.1 above. Appeal only to the total welfare would still entail their removal. Second, the argument in any case works only under very special assumptions. In particular, it requires the assumption that everyone's schedule for the marginal utility for income is the same, that

[handwritten notes in left margin: "cost-benefit approach - equality - maximize welfare - even though agragate maximizing approach aims at maximizing welfare - also equitable"]

[handwritten notes at top of figure: "decreasing marginal utility"]

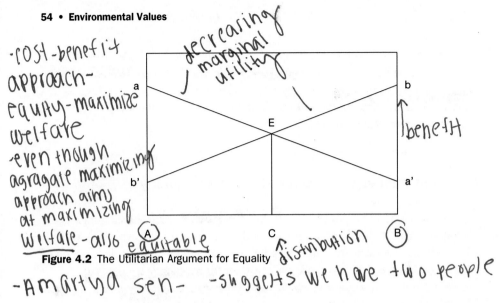

Figure 4.2 The Utilitarian Argument for Equality

[handwritten note: "distribution"]

[handwritten notes: "- Amartya sen - - suggests we have two people"]

[handwritten margin notes: "more you have of something less it will be useful", "-trade-off of distribution", "-agraget welfare is bigget when we equally distribute income"]

individuals get the same level of welfare from each level of income. Without that assumption utilitarianism will have radically inegalitarian outcomes. To use an example of Amartya Sen, assume that Joe gets twice as much utility from a given level of income than Sarah because Sarah is handicapped. Then the way to maximise total well-being would be to give a higher income to Joe than to Sarah. Even if income was divided equally Joe would be better off, and – what is worse – the utilitarian will have to recommend giving more income to Joe (Sen 1997: 15–18).

Figure 4.3 illustrates the difficulty (Sen 1997: 17–18). As with the utilitarian argument for equality from decreasing marginal utility, assume that we are to distribute income between A and B. But assume now not that A and B get the same amount of good or welfare from each unit of income, but rather that A gets twice as much welfare from any level of income as B. The two lines aa′ and bb′ will no longer be mirror images, but rather will be as in Figure 4.3. If we gave equal incomes to each, then A would be better off in welfare terms. His welfare would be given by AaEC, while B's welfare would be given by BbFC. If one wanted to equalise welfare one ought to re-distribute income from A to B. However, if one has the utilitarian goal – to maximise total welfare – then one should do the exact opposite. You should re-distribute income from B to A. The welfare maximising point is at D, in which now the total good that comes to A is AaGD and to B BbGD.

[handwritten note below: "when we equally distribute income"]

[handwritten note at bottom right: "- agrate welfare decline when we depart from caual distribution"]

Economics, efficiency and equality

As we noted in chapters 1 and 2, the main economic decision method employed in environmental decision making is cost-benefit analysis (CBA). It assumes that

[handwritten annotations:]

equality in moral standing - goods+harms that everyone suffer will be added in

- if people do not have same standard of well being - equity is not likely to be an outcome

b - handicapped
- more income to get the same amount of well-being

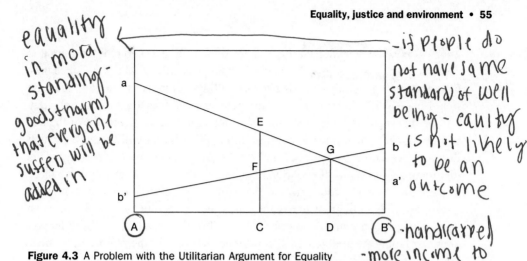

Figure 4.3 A Problem with the Utilitarian Argument for Equality

to improve a person's welfare is to satisfy their preferences, where the strength of preferences can be measured by an individual's willingness to pay for their satisfaction. The benefits of a project are identified by summing the different amounts that affected individuals are willing to pay for the project to proceed, the costs by summing the different amounts affected individuals are willing to pay for the project not to proceed. If benefits outweigh costs then a project is worthwhile. Of a number of projects, the best is that which produces the best ratio of benefits to costs. While CBA can thus be understood in classical utilitarian terms, this is not how many welfare economists normally understand it. Standard economic approaches to environmental decision making do not aim at maximising total welfare as such, as this assumes that we can make comparisons of welfare across different individuals. Rather they seek an *efficient* outcome defined by the Kaldor–Hicks compensation test.

> **Kaldor–Hicks compensation test**: a situation A is an improvement over B if the gains are greater than the losses, so that the gainers could compensate the losers and still be better off.

The optimal decision is that which produces the greatest benefits over costs. Does this offer a more defensible approach compared to classical utilitarianism? The answer is that it does not. Indeed in many ways it is worse. It fails for reasons that we shall now outline.

Willingness to pay

The use of raw willingness to pay monetary measures of the kind employed in standard CBA is incompatible with the principle that equal consideration be given to the interests of all parties affected by a public decision, a principle that classical utilitarians endorsed. First, a person's willingness to pay for a good is dependent upon their budget, on how much they can afford. The poor, since they have less, will generally express a lower willingness to pay for a good than the rich. Hence, while their preferences might be as intense as those of the rich, they will be measured as lower and will count for less. One consequence of the use of raw monetary measures is that the 'efficient' solution will be one in which damages are borne by the poor and the benefits reaped by the wealthy. This distribution of costs to the poor and benefits to the rich will clearly have large implications when it comes to the distribution of environmental benefits and harms. Second, willingness to pay measures cannot capture the interests of those who in principle are unable to offer any bid: future generations and non-human animals. True – the preferences of current generations for the future and for non-humans can be included: the former are captured by 'option values', the latter by 'existence values'. However, this cannot guarantee that proper weight will be given to their interests. If current consumers care but a little, then their interests will count for but a little.

The Kaldor–Hicks compensation test

The Kaldor–Hicks compensation test itself is open to clear distributional objections. These are stated pithily by Sen:

> The compensation principle is either redundant – if the compensation is actually paid then there is a real Pareto improvement and hence no need for the test – or unjustified – it is no consolation to losers, who might include the worst off members of society, to be told that it would be possible to compensate them even though there is no actual intention to do so.
>
> (Sen 1987: 33)

Hypothetical compensations are no compensations at all.

One response is to require that compensation be paid so that we move from a potential to an actual Pareto improvement. However, some environmental goods may be such that there is no sum of money which individuals would accept in compensation for their loss. Consider again the case of the Narmada Dam with which we started. Here is an excerpt from a letter written by a member of the indigenous community from the Narmada Valley in western India, threatened with displacement as a result of the Sardar Sarovar Dam, to the Chief Minister of the state government.

> You tell us to take compensation. What is the state compensating us for? For our land, for our fields, for the trees along our fields. But we don't live

only by this. Are you going to compensate us for our forest? . . . Or are you going to compensate us for our great river – for her fish, her water, for vegetables that grow along her banks, for the joy of living beside her? What is the price of this? . . . How are you compensating us for fields either – we didn't buy this land; our forefathers cleared it and settled here. What price this land? Our gods, the support of those who are our kin – what price do you have for these? Our adivasi (tribal) life – what price do you put on it?

(Mahalia 1994)

Some goods may be such that no price is acceptable. We discuss this point further in the next chapter.

Discounting the future

In addition to these general problems with the Kaldor–Hicks compensation test, CBA also appears to violate the principle of giving equal consideration to the interests of all individuals of whatever generation through its practice of discounting. It weights costs and benefits differently depending on the time at which they occur. It discounts the future. To discount the future is to value the costs and benefits that accrue in the future less than those of the present. By applying a social discount rate, future benefits and costs are converted to current values when aggregating costs and benefits. Since benefits and costs are in cost-benefit analysis measures of preference satisfaction and dissatisfaction, one apparent consequence is that the preferences of future generations weigh less than those of the present. Thus the assumed preferences they might be supposed to have for an absence of toxic waste, expressed in their willingness to pay for that absence or their willingness to accept compensation for its presence, is valued at less than those of current generations. If a preference to avoid some damage is £n, and the annual discount rate is r, then the preference in t years time is $£n/(1+r)t$. Thus, for example, if we assume a preference to avoid toxic damage expressed today at a willingness to pay value of £1,000, then applying a discount rate of 5%, the 'present value' of the same toxic damage occurring in 50 years' time would be $£1,000/(1.05)^{50} = £87.2$. The further into the future, the lower the value.

Discounting has been the subject of much controversy (Parfit 1984: 480–486). In effect, it appears to provide a rationale for displacing environmental damage into the future, since the value placed upon damage felt in the future will be smaller than the same value of current consumption. Some defenders of discounting will argue that differential weighting of costs and benefits over time is actually required *in order to* give equal consideration to interests across generations, in particular to current generations. Three central arguments are offered for using discounting:

1. **Pure time preferences.** Individuals have positive pure time preferences, viz, they are 'impatient'. They prefer benefits now to benefits tomorrow simply in virtue of when they occur. This has a major bearing on the demand structure over time, so the aggregation of individual preferences in cost-benefit analysis must reflect these pure time preferences.
2. **Certainty**. Present benefits are more certain than future benefits. If two benefits are equal, it is rational to give preference to the more certain.
3. **Social opportunity costs**. Any future benefits of a proposal have to be compared to the future benefits that might have accrued had the resources been invested at current rates of interest. Future benefits and costs should, therefore, be discounted by the interest rates expected to prevail over the period of evaluation.

All three of these arguments are, however, flawed. With respect to the first argument, it is far from clear that pure time preferences are rational for individuals (see Pigou 1952: 25; O'Neill 1993: 52–56). However, even if they are, they are a matter for the individual. It is one thing to want your own preferences to be satisfied now rather than later. It is another to assert that your current preferences count more than those of another future individual. The latter involves problems of injustice that do not arise in the former case. With respect to the second argument, there is no reason to assume that degrees of risk and uncertainty are correlated systematically with periods of time into the future in the way that a discount rate assumes. Some harms that will occur in the distant future might be highly predictable; some near-future benefits may be relatively uncertain. With respect to the third argument, appeal to social opportunity costs works only if the goods accrued for the alternative projects are substitutable for those lost in environmental damage. However, for basic environmental goods that are a condition for human life, such as clean air and water, an atmosphere that filters out the sun, there exist no such substitutes. Neither are there for many particular habitats and culturally significant places. The appeal to market interest rates to defend discounts assumes a universal substitutability of goods that may not be met with in practice. We return to the question of substitutability in more detail in chapter 11.

Egalitarian ethics

Neither classical utilitarianism nor the Kaldor–Hicks compensation test that underpins economic decision-making tools, such as CBA, is consistent with the goal of equality. What then should be the response?

Consequentialism without maximisation

One response the utilitarian might make is to retain the consequentialist and welfarist components of the theory, but reject the third purely maximising component. Maximising total welfare is not the only thing that matters. Distribution also matters as an end in itself.

A similar response is often made on behalf of neoclassical approaches that employ efficiency criteria. While efficiency matters, it is not the only thing that matters. The distribution of wealth between different groups also matters. On this view, the Kaldor–Hicks compensation test used in CBA allows the decision maker to determine which of a set of possible projects is the most efficient for society as a whole. Having done so one can then go on to a consideration of its distributional consequences. However, there is a technical difficulty with this defence of the Kaldor–Hicks test. Efficiency cannot be treated as if it was logically independent of distribution – the determination of efficiency already presupposes a given distribution of rights to goods. If you change that distribution of rights you change what is efficient (Martinez-Alier et al. 1999; O'Connor and Muir 1995; Samuels 1981).

However, there is nothing in principle that forces the consequentialist to accept the maximising component of utilitarianism. Nothing in consequentialism rules out putting distributive ends alongside the maximising ends. If one rejects a purely maximising version of consequentialism, what does one replace it with? Two principal candidates have been offered (Parfit 1997).

The priority view

The priority view can be understood as a way of catching the kernel of the utilitarian argument for equality in distribution without the problems associated with marginal utility coupled with maximisation. The basic idea of the priority view is that priority should be given to the interests of the least well-off: the worse-off people are, the more that benefiting them matters. Hence we should distribute goods so that it best improves the condition of the worst off. Thus instead of the maximising component of utilitarianism one might have the following:

> **Priority principle**: Choose the action that produces the greatest improvement in the well-being of the worst off.

The priority principle could be understood simply as a replacement for the maximising principle of utilitarianism, or as a principle that operates in conjunction with maximisation, where there is some further judgement required to deal with conflicts between them.

Adopting the priority principle clearly can make a difference. If one applied that principle to the case of the Narmada Dam, then if the oustees were the worst off, the principle would appear to rule out the project even if total welfare is improved. But if, on the other hand, some of the potential beneficiaries were the worst off, then the project would be justified – though not (solely) because total welfare is improved. The priority view is sometimes claimed to be not strictly speaking egalitarian: redistributions are defended in virtue of the greater urgency of the claims of the less well-off compared to those of the better off, not because equality as such is a good. It might therefore be thought that while the worst off do make greater claims on us there is a problem with the position because it fails to recognise the value of equality as such. Consider the claim that some economic liberals make, that large inequalities of wealth are justified because they improve the condition of the worst off. Large increases in wealth to the already wealthy are justified because they indirectly produce an improvement in the condition of the worst off. There are two lines of criticism one might make of this claim. The first is an empirical criticism, that it is simply not true that inequality benefits the poorest. The second is a normative criticism, that even if it did, there would still be something objectionable about the large inequalities in wealth. If one takes this second line, then it follows that equality has not just an instrumental value, as a means to improving the conditions of the less well-off. Rather, there is something good in equality as such. To hold that position is to defend a form of egalitarianism. However, there are a number of forms that egalitarianism can take.

Telic egalitarianism

One form of egalitarianism is that which combines the value of equality with consequentialism. Rather than argue indirectly for equality, via the priority of the worst off, the consequentialist can assert that equality is an end in itself. If one combines consequentialism with this form of egalitarianism, one arrives at a position sometimes called telic egalitarianism (Parfit 1997):

> **Telic egalitarianism**: We should promote equality because it is a good outcome in itself. What is wrong with inequality is that it is a bad state of affairs as such.

The idea here is that an equal distribution matters in itself, and not just the total amount of happiness. The egalitarian component might replace maximisation, or be put alongside it. In the latter case, in making decisions we need to keep these two distinct dimensions in mind.

A standard objection to telic egalitarianism has been the 'levelling down' objection. One way to realise equality is to reduce the welfare of the better off until it is the same as that of the worse off. This might appear to be prima facie counter-intuitive. For example, in what sense could it be better for everyone to suffer from environmental degradation rather than just some, or for all people to be blind rather than some blind and some not? The telic egalitarian can reply that the position is not committed to saying that equality is the *only* thing that matters in itself, just that it is one of the things that matters. It will allow that where equality is realised through creating a state of affairs in which no one is better off and some are worse off, there is a respect in which the resulting state is better, but this may be outweighed by other considerations, in particular the effect on total welfare. On one dimension of value, that of equality, it is better that everyone suffers from environmental degradation rather than just some, or that everyone is blind rather than some blind and some not. However, says the telic egalitarian, the value of equality here is outweighed by other considerations – the fall in total welfare is too much. What we do in such cases where the total amount of happiness and an equal distribution come into conflict raises further problems that would need to be resolved. But no odd implications need be entailed.

This response may not convince the critic. The critic might want to question whether there is any way in which levelling down could be better – that all suffering environmental degradation could be better than just some, or that everyone being blind could be better than just some. One argument that might be appealed to is that levelling down just doesn't benefit anyone. It only makes some people worse off than they would have been. Hence, how can it be good? The claim that something can only be good if it is good for someone is often called the person-affecting restriction. It is encapsulated in the following slogan:

> **The Slogan**: One situation *cannot* be worse (or better) than another in any respect if there is *no one* for whom it *is* worse (or better) in any respect.
> (Temkin 1993: 256)

Telic egalitarianism violates that slogan. It allows that a state of affairs can be better in one respect without being better for anyone, for example that there is one respect in which it is better for all to suffer environmental degradation rather than just some.

Does it matter that telic egalitarianism violates the person-affecting slogan? The telic egalitarian might respond that there are considerations that give good reasons to reject the slogan anyway. One that is of particular relevance to environmental problems about the distribution of goods over time is what is known as the non-identity problem. (Parfit 1984: ch.16):

> **The non-identity problem**. Consider the choice between two policies, P1 and P2, one of which is more likely to have damaging effects in the future than the other. The choice might be between resource depletion or con- servation, or between high-risk or low-risk energy paths. The policy one

chooses will affect not just the state of well-being of future generations, but who will exist, their identity – population S1 or S2. One of the policies, P1, might produce a much lower quality of life than the other. However, since the population S1 that is produced would not have existed were it not for P1, then, providing their life is worth living, they cannot be said to have been harmed, since they are not worse off than they would have been had they not existed. There is no specific person who is wronged or harmed. Hence, if we accept the person-affecting slogan, there is nothing wrong with choosing the more environmentally damaging policy over the lesser.

However, given this implication there are good reasons for arguing in the opposite direction. Clearly there is something wrong with choosing a policy P1 that issues in a lower quality of life, even if there is no particular person who is made worse off by it. Hence it is the person-affecting slogan that has to go. If it goes then the telic egalitarian can argue that it loses any force against the egalitarian position. There is something good about equality even for those cases in which equality benefits no particular person.

Neither the levelling-down objection nor the appeal to person-affecting constraints are decisive against telic egalitarianism. However, there is something problematic about telic egalitarianism which the objections point towards, but do not really pin down. The problem with telic egalitarianism is that it doesn't appear to capture the kind of reasons that are normally offered for thinking that equality is a good thing. Consider for example the following scenario: There are two islands independent of each other and unknown to each other. One is a desert that can grow few crops and the population suffers in virtue of this fact. The other is an agricultural paradise, in which crops grow freely, and the islanders live well. A natural disaster hits the second island reducing it, too, to a desert. Equality between the islanders has been achieved. However, it would be odd to think that in any respect the disaster served to make the world a better place. The mere existence of equality of condition, in abstraction from the kinds of relations and interactions individuals and groups have to each other, has no value. The value of equality doesn't appear to be captured by the telic egalitarian.

Deontological responses

Consider again the putative counter-examples to utilitarianism of the kind illustrated by the Narmada Dam case. The telic egalitarian responds by keeping the welfarism and consequentialism of classical utilitarianism, but rejecting the maximising component. A second response might be to reject the consequentialist assumptions of utilitarianism. What is wrong with the reasoning in the Narmada Dam case is that it fails to consider the possible *act* of injustice against the oustees that it involves, even if the overall well-being is improved.

One consideration that might be invoked in favour of that response is a broadly Kantian view of moral standing that rejects the account of equality in moral standing that utilitarianism offers. On the utilitarian view all sentient beings have equality of standing in the sense that we should give equal consideration to their interests in appraising which policy maximises welfare. On the Kantian view to have moral standing is to possess certain interests which cannot be over-ridden for the purposes of maximising welfare. Equality in moral standing on this view involves recognition that a being is an end in itself, and to be treated as such is not to be used merely as a means to other ends, including the maximisation of the general welfare. As Rawls puts it: 'Each person has an inviolability founded on justice that even the welfare of society as a whole cannot override' (Rawls 1972: 3). Utilitarianism fails to acknowledge the separateness of persons.

Rawls's theory of justice remains the most influential of recent deontological accounts of justice. Rawls's theory shares with utilitarianism the assumption that justice involves impartiality, but impartiality is compatible with recognition of the separateness of persons. The device of agreement in the original position is taken to capture the impartiality of justice without sacrificing separateness. The principles of justice are those that rational individuals would agree upon in the 'original' position, that is the hypothetical situation in which mutually dis-interested individuals choose principles behind a 'veil of ignorance' regarding their social position, characteristics and conception of the good. Since they are behind a veil of ignorance they cannot choose principles which favour their individual interests. Thus they represent people who choose principles as though in ignorance of how their choice would affect themselves personally, in other words, impartially. What do individuals know in the original position? They know that their society is subject to the circumstances of justice (limited altruism and moderate scarcity) and 'general facts about human society'. While they do not know their particular conception of the good, they do have knowledge of 'primary goods', understood as those things which it is supposed a rational person would want whatever else she wants: 'Whatever one's systems of ends, primary goods are necessary means . . . While the persons in the original position do not know their conception of the good, they do know . . . that they prefer more rather than less primary goods' (Rawls 1972: 93).

What principles would individuals choose in such conditions? Rawls argues that agents under such conditions would arrive at two basic principles:

P1. Each person has an equal right to a fully adequate scheme of equal basic liberties which is compatible with a similar scheme of liberties for all.

P2. Social and economic inequalities are to satisfy two conditions. First, they must be attached to offices and positions open to all under conditions of fair equality of opportunity [the opportunity principle];

and second they must be to the greatest benefit of the least advantaged members of society [the difference principle].

These principles are lexically ordered: the satisfaction of the first principle has priority over the satisfaction of the second and the first part of the second is prior to the second part of the second.

Rawls's theory of justice is egalitarian about rights, basic liberties and opportunities. To fail to respect persons' equal rights to liberty is wrong since it involves an act of injustice. However, it is not egalitarian about the distribution of other primary goods. About other primary goods, Rawls adopts a deontological version of the priority principle. Inequalities are justified if they benefit the worst off group. That version of the priority view is open to much the same problems as the consequentialist version discussed above. Equality often appears to matter in itself, not just as a way of improving the condition of the worst off.

Why should equality matter? The telic egalitarian claims that equality is simply a good state of affairs as such. We noted above that this view doesn't appear to capture the fact that the value of equality appears to have something to do with the kinds of social relations individuals and groups have to one another. One response to that objection might be to combine egalitarianism with a deontological ethic (Parfit 1997).

> **Deontic egalitarianism**: We should aim at equality because to do so is to perform the right or just action. What is objectionable about inequality is that it involves wrong-doing or acts of injustice.

Deontic egalitarianism need not as such be open to the levelling down objection to telic egalitarianism. Nor is it open to the two-island counter-example we outlined at the end of the last section – there was no act of injustice involved in that case. It does allow that the value of equality may be tied to the various relations that individuals have to each other. In introducing justice one might argue that equality is good because it is a constitutive condition of social relations within a just community.

Community, character and equality

However, while considerations of justice might form part of the grounds for seeking equality, it is not clear that this appeal to community need be couched purely in the language of justice. Certainly, appeals to justice do not capture everything that informed the ideal of equality. Consider for example the following characterisation of the ideal by George Orwell:

> Up here in Aragon one was among tens of thousands of people, mainly though not entirely of working-class origin, all living at the same level and

mingling on terms of equality. In theory it was perfect equality, and even in practice it was not far from it. There is a sense in which it would be true to say that one was experiencing a foretaste of Socialism, by which I mean that the prevailing mental atmosphere was that of socialism. Many of the normal motives of civilized life – snobbishness, money grubbing, fear of the boss, etc. – had simply ceased to exist. The ordinary class-division of society had disappeared to an extent that was almost unthinkable in the money-tainted air of England; there was no one there except the peasants and ourselves, and no one owned anyone else as his master.

(Orwell 1966: 101–102)

The ideal of equality here is tied to a wider set of virtues than that of justice. Indeed justice in a narrow sense is not mentioned. However, in his account of the community of Aragon, Orwell does capture some of the wider grounds of the appeal of equality. If one considers why inequality may be objectionable even where the condition of the least well-off is improved, the answer is not just that justice is not done, but also that inequality is a constitutive condition of other vices such as dependence, humiliation, snobbery, servility and sycophancy. Central to the history of egalitarian thought is the argument that equality is a condition for a certain kind of community and human character. The ideal of equality is tied to the creation of a community in which certain forms of power, exploitation and humiliation are eliminated and solidarity and fellowship fostered (Miller 1997; O'Neill 2001; Tawney 1964). It might be that an egalitarian community requires departures from strict justice, but that it is none the worse for that: justice is one virtue among others, and we are sometimes willing to allow it to be subordinate to others – for example, generosity or mercy. So, as an alternative, consider this:

Virtues/community-based egalitarianism: We should aim at equality because it is a constitutive condition of certain social virtues. What is objectionable about inequality is that it engenders social vices.

To defend equality by appeal to the relationships and virtues of character that it fosters is not necessarily to take equality to have merely instrumental value in the sense of being an external means to a distinct end: equality and the mutual recognition of equality are partly constitutive of many of the virtues and relationships to which appeal is made (Norman 1997: 241).

This virtue-based egalitarianism does stand in contrast to liberal theories of justice of the kind that Rawls offers. The Rawlsian theory is one of many attempts by liberals to express the idea that public institutions should be neutral between different conceptions of what a good life is like and the different communities and practices through which these are lived. For the liberal it is not the job of public institutions to promote a particular kind of life. That is up to individuals to choose for themselves. What we require is a set of institutions that are governed by principles of justice that are themselves neutral between those different conceptions. To capture that idea Rawls assumes that in the original position individuals only have knowledge of what he calls 'primary goods' – that is, those goods that are necessary to pursue any conception of the good: we do not know what our specific conception of a good life will be like. In contrast, the virtues-based account of equality does make some substantive assumptions about what the good life for humans consists in. It assumes that there are excellences in the human character and relations to be promoted. Moreover, it makes claims about the nature of a good community.

When it comes to environmental arguments, appeals to our understanding of the kind of community to which we belong do appear important. Some environmental goods are clearly primary goods in Rawls's sense. Indeed the life support functions of the environment offer excellent examples of what primary goods are supposed to be: they are goods without which we could not pursue any conception of the good life. Without minimal standards in the state of air and water, for example, a human life cannot be lived at all. However, many environmental concerns are not of this life support kind. As we noted in the first chapter, we do not only live from nature, but also in it and with it. Environments form a central component in the identities of individuals. Particular places matter to individuals in virtue of embodying their history and cultural identities and this is why their loss is felt so acutely. The loss of forests, the damming of rivers with the subsequent flooding of villages and their natural setting, the disappearance of particular economically and biologically insignificant places, 'natural' and 'urban', the displacement of populations to make 'nature reserves' – all matter because the environments embody in a physical way the identity of individuals and the communities to which they belong. Their loss involves the loss of a way of life. Individuals feel a loss of something integral to their lives. To say this is not to deny the desirability of change, even radical change: it is to say that the nature of such changes and of the transition to new ways of life matters. Environments are not just of instrumental value, or a physical precondition of human life: individuals' identities, their sense of who they are, is partly constituted by their sense of belonging to particular places. These features of some environmental goods make them difficult to fit into the picture of neutral public institutions offered by Rawls and other liberals. To say that these considerations should matter in public deliberation and that public decisions should foster and develop particular kinds of community is to reject the ideal of neutral public institutions. It should be added that to say this is not necessarily to reject liberalism, for there are versions of liberalism that themselves reject neutrality (Raz 1986).

This virtues- or community-based account will also point to a different account of the relations between generations, which will be much more about sustaining and developing communities over time than it is about contracts between disinterested strangers, or impersonal concerns for others. We have a responsibility to attempt, as far as is possible, to ensure that future generations do belong to a community with ourselves – that they are capable, for example, of appreciating works of science and art, the goods of the non-human environment, and the worth of the embodiments of human skills, and are capable of contributing to these goods. This is an obligation not only to future generations, but also to those of the past, so that their achievements continue to be both appreciated and extended, and lastly to the present, ourselves. This is not to deny that we have obligations to future strangers. It is to note that intergenerational justice cannot be adequately modelled on an account in which we treat each other and ourselves as strangers. (See O'Neill 1993: ch.3, and de-Shalit 1995 for a development of this position.) We return to this point in more detail in part three.

Equality of what?

The different forms of egalitarianism we have considered so far in this chapter can be understood as different answers to the question 'Why equality?' The answers to that question are closely related to answers to a second question concerning equality: 'Equality of what?' In considering distribution, what is it we are distributing? There are a variety of different answers to that question in the literature: equality of welfare and of resources (Dworkin 1981); equality of liberty and in the distribution of primary goods (Rawls 1972); equality of opportunities of welfare (Arneson 1989); equality of access to advantage (G. A. Cohen 1989); equality of 'capabilities to functionings' (Sen 1980, 1997); equality in goods that fulfil objective interests and needs. Which answer one gives has important implications for the discussion of sustainability to which we will return later. If sustainability is about equity in distribution over generations, then it raises the same question as to what it is we are supposed to be distributing equally. The question 'Equality of what?' is directly related to the question common among environmental activists and advocates, 'Sustainability of what?' One idea is that the goal of environmental protection should be to produce conditions that are environmentally sustainable for future generations so that they can enjoy as good (or better) environmental quality as we have experienced. And while the issue is not always raised in this form, one way of understanding the debates between the various conceptions of environmental sustainability is that they differ on what we are supposed to be distributing: for example, whether we should aim at sustaining (i) levels of well-being, be this understood in terms of preference-satisfaction (Pearce 1993: 48) or objective goods, (ii) options or opportunities for welfare satisfaction (Barry 1997), (iii) resources and environmental capacities (Jacobs 1995), or (iv) capacities to meet needs or objective interests (World Commission on Environment and Development 1987). We will return to these questions in chapter 11.

However, to the extent that equality is concerned with the nature of the communities in which we live, many of these answers to the question 'Equality of what?' tend to have too tight a focus on external objects or states of affairs to which individuals might have different levels of access. While distribution in this sense is important, it tends to miss other significant dimensions of equality. Equality is concerned with the nature of the social relationships between individuals. It is concerned with issues of power, of relationships of super-ordination and subordination, and with issues of recognition, of who has standing in a community and what activities and attributes are acknowledged as having worth. The poor do not just lack the means to satisfy their needs, important as this is. They also tend to be powerless and socially invisible, their activities not given proper recognition. Both of these dimensions matter in the environmental sphere. They are often missed in discussions of the environment because of the focus in environmental philosophy on issues of justice and ethical obligation to non-humans and future generations. In one way this focus is understandable, since these are the issues highlighted by specifically environ-mental problems. However, it also has some unfortunate consequences. The focus has tended to reinforce the tendency to concentrate on the distribution of goods and harms, at the expense of questions about the distribution of power and recognition. At the same time, there has been a lack of acknowledgement of the way that environmental problems raise issues of equity within current generations of humans which are just as pressing in a practical sense as those that have dominated ethical debate in environmental philosophy.

Consider some of the problems with which we opened this chapter: the siting of toxic waste dumps and incinerators, of open cast mines, roads, runways and power stations; the differential effects of deforestation across gender and ethnic groups; the creation of new systems of intellectual property rights on genetic resources. The focus on environmental distribution within current generations of humans highlights dimensions of inequality that have tended to be overlooked in much of the ethical debate in environmental philosophy. The question 'Equality of what?' has answers that refer not just to external objects or states of affairs to which individuals might have different levels of access, but also to the nature of the social relationships between individuals and groups: with issues of power between groups, of domination and subordination, and with issues of recognition concerning which actors and activities are socially acknowledged as having worth. With respect to both future generations and the non-human world, the problem of the distribution of power tends to be secondary, since of necessity the power of decision does not lie with them, for all they might thwart decisions made. However, within current generations, environmental conflicts are often between those who have the economic, political and social power to shape economic projects and their outcome, and those who lack that power and who are driven to resist their outcome. The conflicts around what Martinez-Alier has called the 'environmentalism of the poor' (Guha and Martinez-Alier 1997) are of this nature. Similarly, problems of social recognition are also brought to the fore. In particular, many recent environmental disputes concern not just the social invisibility of the poor, but more specifically, the lack of social recognition of forms of labour such as domestic labour and subsistence agriculture that fall outside the market realm. Within a market system, those forms of labour which have been central to sustaining human life and agricultural biodiversity tend to go unrecognised and unvalued (Martinez-Alier 1997 and Shiva 1992). Environmental conflicts concern not just the distribution of environmental harms and goods, but also of power and recognition.

5 Value pluralism, value commensurability and environmental choice

In chapter 2 we noted that classical utilitarianism, the view that the best action is that which produces the greatest improvement in total happiness, makes three distinct claims:

1. it is **welfarist** – the only thing that is good in itself and not just as a means to another good is the happiness or well-being of individuals;
2. it is **consequentialist** – whether an action is right or wrong is determined solely by its consequences;
3. it is an aggregative **maximising** approach – we choose the action that produces the greatest total amount of happiness or well-being.

In the previous three chapters we have examined the problems raised by each of these three claims. In this chapter we examine two additional general assumptions that we can now understand as underpinning the first and third components of the theory – **value monism** and **value commensurability**. Both assumptions matter for environmental issues in two distinct ways. First, in environmental philosophy there has been a major conflict between those who think that any coherent and rational approach to environmental choice requires both monism and commensurability of values, and those who allow that rational choices are possible without those assumptions (Stone 1988; Callicott 1990; Wenz 1993; Light 2003). Second, in environmental policy practice there is an assumption, often implicit but sometimes made explicit, that rational decision making requires value commensurability. In particular, the dominant economic approaches assume that to make rational choices we must adopt some common measure of value, and that money provides that measure.

The first component of classical utilitarianism, welfarism, claims that there is only one thing that is of value in itself and not as a means to another end, namely happiness. In making this claim it looks as if it is committed to a version of value monism. (Whether it is really so committed is a question we shall return to later.) What is value monism? Value monism is the claim that there is only one intrinsically valuable property or entity which is valuable in itself, and that other values are reducible to this value. Thus Bentham, for example, who offers a **hedonist** version of welfarism that identifies happiness with pleasure and the absence of pain, holds the ultimate single value to be pleasure. However, there are other forms of value monism which start with a different single value. Thus some Kantians offer respect for human dignity as the ultimate value. Value monism contrasts with value pluralism, the view that there are a number of distinct intrinsically valuable properties, such as autonomy, knowledge, justice, equality, beauty, which are not reducible to each other nor to some other ultimate value such as pleasure.

The third component of classical utilitarianism, the maximising component, makes a second assumption. It assumes the existence of a common **measure** of value through which different options can be compared in order to ascertain which produces the greatest total amount of value. To make this assumption is to be committed to **value commensurability** – the claim that there exists a common measure of value through which different options or states of affair can be ordered. What kind of scale of measurement do we require? Classically, utilitarians assumed the existence of a *cardinal* scale, roughly speaking, a scale that provides information on precisely how much value different options offer. More recent economists and philosophers within the utilitarian tradition have tended to argue that we only need an *ordinal* scale, that is, a scale that simply ranks the value different options offer – 1st, 2nd, 3rd and so on – without assigning them any specific value on how much they differ. (To understand the difference, consider comparing three sprint athletes, Jones, Ward and Simpson, and the assessment of their respective merits in terms of their speed over 100m. A cardinal scale will assign a precise speed to each athlete, 9.2 seconds to Jones, 9.4 seconds to Ward, 10 seconds to Simpson. An ordinal scale will simply rank them in terms of the speed, Jones as fastest, Ward as second fastest, Simpson as slowest.)

Part of the promise of utilitarianism, and standard economic decision-making tools such as cost-benefit analysis, is that they offer a procedure for resolving value conflicts through the employment of a common measure which allows the losses and gains within each option to be aggregated, and then the total value offered by each option to be compared. It promises the possibility of reducing social choice to a matter of a calculus – a method by which anyone, given a set of data about the outcomes of alternative actions, can work out mechanically which outcome is best.

Given this ambition one attraction of at least some forms of value monism is that they offer the possibility of arriving at cardinal measures of value. The classical example of this move is to be found in Bentham's version of utilitarianism. In Bentham's case, pleasure is the only ultimate value. Pleasure is taken to be a homogeneous commodity, of which people's experiences contain different, measurable, amounts. Pains can be measured on the same scale as pleasures, only as negative quantities. Thus it is possible to aggregate the pleasures and pains experienced by each person, and then aggregate the pleasures and pains of a number of people affected by some action, in this way arriving at a single result which gives the total utility of that action.

Mill's insistence that pleasures differ in quality and not only in quantity is in effect a criticism of Bentham: if he is right there can be no arithmetical calculus of pleasures of the kind that Bentham offers. But he suggests an alternative method by which pleasures can be compared: pleasant experiences are desired, and some are desired more than others. Two pleasures as disparate as poetry and pushpin (or some modern equivalent, say playing a slot-machine) may still be compared, and one rated more valuable than the other. The criterion is whether it is preferred, or desired *more*.

Contemporary utilitarians, as we noted in chapter 2, tend to assess consequences in terms of preference satisfaction. And in this they have been followed by most welfare economists. Part of the attraction for the economist lies in the possibility that it opens up of employing the 'measuring rod of money'. The intensity of a person's preferences for different goods is taken to be measurable by how much they are willing to pay at the margin to acquire the good or how much they are willing to accept for its loss.

Value monism

Bentham and Mill are both value monists. There is only one source or standard of value, namely pleasure, through which we can compare options. Why should it be thought there must be but one standard of value? One response to the question is to argue that a single standard of value is required to compare options if there is to be rational decision making at all. A classic statement of the position is the following argument from J. S. Mill:

> There must be some standard to determine the goodness and badness, absolute and comparative, of ends, or objects of desires. And whatever that standard is, there can be but one; for if there were several ultimate principles of conduct, the same conduct might be approved of by one of those principles and condemned by another; and there would be needed some more general principle, as umpire between them.
>
> (Mill 1884, bk 6, ch.12, § 7)

If we have several ultimate principles that embody different standards of value, then they might specify conflicting actions. If we are to avoid making an arbitrary decision, or no decision at all, then we need some way of adjudicating between different standards when they conflict. Hence we will need another general principle that will act as an adjudicator between conflicting principles. If there is not to be yet another conflict, this umpiring principle must specify a single standard of value.

There are several problems with this argument. Two are worth noting here.

1. Mill claims that if there are a number of different standards that come into conflict then there must be a general umpiring principle to resolve the conflict. However, it does not follow that there must be just one standard to determine the goodness or badness of different ends and objects – that the only umpire is a single measure of value. An alternative possibility is to have many standards of value, v_1, v_2 ... v_n, and some ordering principle for determining which takes precedence over others, an umpiring *rule*. As we noted in the previous chapter, Rawls introduces an ordering principle of this kind, what he calls a lexicographic ordering amongst values v_1, v_2 ... v_n, such that v_2 comes into play only after v_1 is satisfied, and in general any further standard of value enters into consideration only after the previous value has been satisfied (Rawls 1972: 42ff and 61ff). As we further noted, the notion of rights in moral argument also introduces a lexicographic ordering of this kind. Rights are trumps in moral arguments (Dworkin 1977: xi). Rights win against any other values although they resolve disputes only where trumps play another suit of values. There is also empirical evidence in the environmental sphere, for example, when people are questioned about their valuation of woodland and forest protection, that some respondents are expressing 'lexicographic' preferences when they refuse to entertain a trade-off between environment and income (Spash and Hanley 1994).

This is not to say there are not problems with the lexicographic approach. Strict lexicographical orderings of different values are notoriously implausible at the edges. For example, the idea that environmental goods have some kind of lexicographic priority over other constituents of human welfare looks dubious. Is it plausible to assert that some environmental good has to be met in the face of any other claim? If it is not, the critic can argue that all that is really going on here is that at certain levels in the satisfaction of one set of goods – say material welfare – the weight on environmental goods simply becomes very high: it would take a huge increase in material welfare to compensate for a loss in environmental quality. There is no absolute priority principle of a lexicographic kind at work at all. In response, it is possible for a defender of the lexicographic approach to modify the orderings, which would allow for more moderate positions such as a refusal to trade-off, say, liberty or environmental goods, once some minimum level of welfare is realised.

2. In fact the appeal to an umpiring rule between plural values still shares with Mill's position the assumption that a rational resolution of practical conflicts requires a single principle of adjudication. The disagreement concerns whether it be a single supervalue to which all others can be reduced, or a priority rule amongst plural values that are not so reducible. There is, however, a flaw in Mill's argument for the claim that there must be a general umpiring principle of either kind. It might be true that for any conflict between values there must be some way of resolving that conflict. It does not follow that we need to come up

with a single value or umpiring rule that will resolve every single conflict. To suppose that it does follow is to make an inferential error known in logic texts as a shift in the scope of a quantifier: To see what is wrong, compare the inference from 'for any person there is someone who is that person's parent' to 'there is someone who is the parent of every person'. We consider the implications of this point further below.

Value pluralism

Value pluralism in the sense we use it here is the view that there are a number of distinct values, such as autonomy, knowledge, justice, equality, beauty, etc., which are not reducible to each other nor to some other ultimate value such as pleasure. One thought in favour of value pluralism lies in the very richness of our evaluative vocabulary. Consider the diversity of evaluative concepts that we employ in the appraisal of our environments, and the variety of practices that inform our relationships to our environments that this reflects. From the biological and ecological sciences come concepts such as biodiversity, species richness, integrity, fragility, health. From aesthetic traditions come a variety of concepts that we call upon in our appreciation of natural beauty: tones of colour, such as the contrasting browns and reds of autumn, the subtle shifts in shades of green in spring, the dappled sunlight in woodlands; the forms and shapes of nature such as the ruggedness of mountains, the gentleness of hills, the landscapes shaped by stone wall and terrace; the sounds such as birdsong and river over rocks; and textures such as the roughness of gritstone and the sharp and smooth of limestone. There are also the moods of nature, the violent wildness of a storm, the serene lakeside, the force of a waterfall. And as a place, a location might be valued for being evocative of the past. We use then a rich vocabulary to appraise the environments we live in, from and with. The idea that this could be reduced to a single value looks implausible.

One obvious response to this thought is that when all is said and done there is still only a single value in play here. The different values all make a contribution to human well-being, or more widely the well-being of individual sentient beings. Well-being remains the ultimate value. However, even if that is accepted, it is not clear that a commitment to value monism follows. And here we make the reassessment we promised earlier in our discussion of value monism and welfarism. For some apparently monist views turn out to be pluralist, if one digs a little deeper.

Consider again the welfarist assumption of utilitarianism. The welfarist says that only one thing matters, the well-being of individuals. As noted above, this looks like a monist assumption, and in the hands of a writer like Bentham it was so understood. However, some versions of welfarism are themselves pluralistic when unpacked. For example, one might think that only the well-being of individuals is of value in itself, but hold that there are a number of different components of well-being that are not reducible to each other. Consider the objective list views of welfare of the kind outlined in chapter 2. The objective list accounts of well-being typically assume that there are a number of components of well-being – autonomy, knowledge, personal relationships with friends and family, and so on. Each is valuable in itself and not reducible to any other. Welfare or well-being are covering terms for the various valuable components of life. Welfarism on this account would still be pluralist. Indeed, it is worth noting that even hedonism can turn out to be pluralist when considered a little more deeply. For, contrary to what is commonly assumed, pleasure is not a single value. A long cool drink at the end of a long hot walk, a conversation with a good friend, watching buzzards wheeling in the sky, achieving a long-time ambition to run a marathon within a particular time – all are a source of pleasure, but they do not share in some common property of pleasurableness which we can add and subtract. There is a prima facie case for value pluralism even from within a hedonist perspective.

Trading-off values

If value pluralism is true, does it follow that there is no single measure of value, that value commensurability also falls? Not necessarily. We noted above that one way of getting value commensurability is, like Bentham, to assume value monism. However, this is not the only way. Value commensurability is compatible with at least some forms of pluralism, if one assumes that different values can be *traded off* with each other.

What is it to say that values can be traded off against each other? The answer runs something as follows. There are a variety of ultimate values, but we can compare those values and say that a loss in one dimension of value is equal to a gain in another. In cases where different values conflict, one has then not an umpiring principle of the kind that Mill outlines, nor a lexicographic ordering of value, but rather a trade-off schedule that says that a loss of so much in one value is compensated for by a gain of so much in another. In making choices or expressing preferences between different options which involve losses and gains in different values, we are implicitly trading off different values.

If one assumes that in making choices we do implicitly have a schedule for trading off different values then the claim that there is a universal *measure* of value, even if not a single value, looks as if it is back on the cards. Value commensurability may be possible. What we need is a universal currency for that trade, some measuring rod which we can use to measure the different rates at which losses and gains in different dimensions of value evidence themselves, and then put them on a common scale. And a thought that moves the economist in the utilitarian tradition is that there is such a universal measuring rod – money – which is used in the market place to bring different dimensions of value under a common metric. On this view, money is not an ultimate value in itself, nor is it a measure of some other value like pleasure. Rather, it is simply a metric which we can use to put on a common scale the relative importance of different values. Thus consider the following observation by Pearce et al. on the use of physical descriptions of environmental goods as against monetary valuations:

> Physical accounts *are* useful in answering ecological questions of interest and in linking environment to economy . . . However, physical accounts are limited because they lack a common unit of measurement and it is not possible to gauge their importance relative to each other and non-environmental goods and services.
>
> (Pearce et al. 1989: 115)

The Pearce report assumes that to make a rational non-arbitrary choice between options there must exist some 'common unit of measurement' through which the relative importance of different goods can be ascertained. Money provides that unit of measurement and cost-benefit analysis the method of using that unit in decision making: 'CBA is the only [approach] which explicitly makes the effort to compare like with like using a single measuring rod of benefits and costs, money' (Pearce et al. 1989: 57). The problem for environmentalists, on this view, is to extend the 'measuring rod' of monetary units to include those environmental goods which are at present unpriced. Hence the development of methods aimed at the money estimates of the 'marginal benefits' of environmental protection.

Thus, consider the example of forestry management in the UK. The context is one of conflicting values. There exist conflicts, for example, between different biodiversity objectives. Increasing the diversity of native tree species in forests is in conflict with the aim of protecting the native species of red squirrel which fares better than the immigrant grey squirrel in conifer plantations; increasing the diversity of native tree species can conflict also with the protection of the goshawk which flourishes in spruce plantations. These competing biodiversity considerations themselves conflict with others: landscape considerations may dictate a particular mix of trees; the use value of forests as a timber resource will suggest different priorities; woodland may have specific historical and cultural meanings as a place for a particular community that would suggest something different again. How are such conflicts between different objectives to be settled? One way that is widely used is to have an argument about it. Botanists, ornithologists, zoologists, landscape managers, members of a local community, timber companies, saw mill owners, unions, farmers, representatives of various recreationalists all argue their corner. Out of this a decision is supposed to emerge.

There are clearly problems with that approach. As we noted in earlier chapters, the decision that will emerge will reflect not necessarily the soundness of the arguments of the different groups but rather their relative power with respect to each other. There will be interested parties who are not represented. In particular, decisions will involve the use of resources that might have been used for non-forestry schemes – for schools, health services, and the like: any decision has what economists call 'opportunity costs'.

The worry runs that these argumentative procedures are not how it should be done if we are to do things in an ideally rational way. What the utilitarian approach promises is the replacement of this fairly messy looking procedure by something that is taken to be more rational. Ideally, we should employ a common measure through which we can trade off different objectives in an impartial and objective manner. Given the existence of competing values and objectives – biodiversity, landscape, timber, cultural meanings, historical and scientific values, and so on – resolution requires some common measure of comparison for giving each its due against the other. Certain economists argue that monetary price is the best measure for making those trade-offs. The use of the measure does not entail that there is only one thing of value, in particular that only money is of value. Rather, money serves as a measure of the exchange rate between different values.

Constitutive incommensurabilities

Can values be traded like this? There are problems. An initial point to note, to continue using the trading analogy, is that the currency of some values may not be convertible into the currency of others. Some exchanges are blocked. The point is of particular significance for the use of money measures. There are many values which simply cannot be converted into a monetary equivalent. The values are such that losses and gains cannot be caught in monetary terms. The problem here is that money is not just a 'measuring rod'. There is a social meaning invested in acts of exchange. Correspondingly, certain social relations and evaluative commitments are constituted by a refusal to put a price on them (Raz 1986: 345ff; O'Neill 1993: 118–122). If I care about something, then one way of expressing that care is by refusing to put a price on it. Older willingness to pay surveys are aware of this. Consider the following from Herodotus's histories:

> When Darius was king of the Persian empire, he summoned the Greeks who were at his court and asked them how much money it would take for them to eat the corpses of their fathers. They responded they would not do it for any price. Afterwards, Darius summoned some Indians called Kallatiai who do eat their parents and asked in the presence of the Greeks

... for what price they would agree to cremate their dead fathers. They cried out loudly and told him not to blaspheme.

(Herodotus Histories 3.38)

Darius's 'willingness to accept' survey, unlike the willingness to pay surveys of his modern economic counterpart, aims to elicit protest bids. The story would have been somewhat ruined if the Kallatiai had responded by putting in a realistic price. The reason why Darius elicits the protests is to reveal the commitments of the individuals involved. One exhibits commitment to some good, here one's dead kin, by refusing to place a price upon it. In contrast, the modern economist begins by ignoring all protest bids: these, together with strategic responses, are laundered out of the process, to leave us with just those considered relevant to a calculation of the welfare benefits and costs of the project. Part of the problem here is with the view of monetary prices that economists in both the neoclassical and Austrian traditions assume. Monetary transactions are not exercises in the use of a measuring rod. They are social acts which have a social meaning.

Certain kinds of social relations and evaluative commitments are constituted by particular kinds of shared understanding which are such that they are incompatible with market relations. Social loyalties, for example, to friends and to family, are constituted by a refusal to treat them as commodities that can be bought or sold. Given what love and friendship are, and given what market exchanges are, one cannot buy love or friendship. To believe one could would be to misunderstand those very relationships. To accept a price is an act of betrayal. To offer a price is an act of bribery. Similarly, ethical value commitments are also characterised by a refusal to trade. Such refusals are found in protests against requests to price environmental goods. Environmental goods are often expressive of social relations between generations. They embody, in particular places, our relation to the past and future of communities to which we belong and the values that are thought to matter to those communities. And it is that which in part activates the protests against the demand to express concern for nature in monetary terms, including protests from some who may have actually responded 'legitimately' to the survey. Consider the following two protests. The first is from a respondent to a contingent valuation survey:

> it's a totally disgusting idea, putting a price on nature. You can't put a price on the environment. You can't put a price on what you're going to leave for your children's children . . . It's a heritage. It's not an open cattle market.

(Burgess et al. 1995)

The second is a response to an actual request to price an environment. How much would you be willing to accept in compensation for the loss of your home occasioned by a dam project? Here again is the excerpt quoted earlier from the letter of an inhabitant of the Narmada Valley in western India, threatened with displacement as a result of the Sardar Sarovar Dam, written to the Chief Minister of the state government.

> You tell us to take compensation. What is the state compensating us for? For our land, for our fields, for the trees along our fields. But we don't live only by this. Are you going to compensate us for our forest? . . . Or are you going to compensate us for our great river – for her fish, her water, for vegetables that grow along her banks, for the joy of living beside her? What is the price of this? . . . How are you compensating us for fields either – we didn't buy this land; our forefathers cleared it and settled here. What price this land? Our gods, the support of those who are our kin – what price do you have for these? Our adivasi (tribal) life – what price do you put on it?'

> (Mahalia 1994)

The point here is that an environment matters because it expresses a particular set of relations to one's community that would be betrayed if a price were accepted for it. The treatment of the natural world is expressive of one's attitude to those who passed the land on to you and to those who will follow you. Money is not a neutral measuring rod for comparing the losses and gains in different values. Values cannot all be caught within a monetary currency.

Value pluralism, consequentialism, and the alternatives

One problem with the metaphor of trading off values is that some values may not be convertible into other currencies of value, and in particular into the currency of money. A second problem is that the standard interpretation of the metaphor suggests a particular consequentialist framing of choices. What such a framing assumes is that we are trying to produce a state of affairs with the greatest amount of value, and we do this by comparing the gains and losses of different dimensions of value, trading these off until we arrive at a result that produces the greatest gains in values over losses in values. The account is pluralist about values – there are a variety of different values – but by exchanging gains and losses in these distinct dimensions of value, we can still arrive at some notion of the outcome with the highest total value. Other approaches to ethics, deontological- and virtue-based approaches, do not lend themselves to the trading metaphor in this sense.

Both deontological and virtue approaches to ethics can take either monist forms or pluralist forms. The deontologist can take a monist position, as some Kantians do – that there is one basic obligation, for example, that we treat persons as ends in themselves, and that other obligations are derivatives of this obligation.

Alternatively it can take a pluralist form. The pluralist will hold (i) that there are a variety of basic obligations, say obligations of justice, obligations to improve the well-being of others, obligations not to injure others, obligations to develop one's own projects, obligations that arise from special relations (for example, of parents to children), obligations that arise from previous acts, say of making a promise, or making right a previous wrong, or from gratitude, and so on; and (ii) that these obligations are not reducible to each other or some other ultimate principle. David Ross defended a view of this kind (Ross 1930). If one holds this position then one will allow that these obligations can sometimes come into conflict with each other. Likewise virtue ethics can take monist and pluralist forms. One might think there is just one ultimate virtue, such as intelligence, and that all other virtues – courage, justice, generosity, autonomy, kindness, and so on – are ultimately different ways of exhibiting that virtue. (Socrates is thought to have held such a view.) Or one can hold that there are many basic virtues that are not reducible to each other or some other ultimate virtue. Given pluralism, there is again the possibility that these conflict.

However, neither deontological nor virtues approaches will assume that we resolve value conflicts in a consequentialist manner by trading off the gains and losses across different values and working out which produces the greatest amount of value. From a deontological point of view, resolving a value conflict is rather a matter of which obligation has the stronger pull on an agent in a particular context. Consider a parent who spends time with her children on holiday, which she could spend acting for an environmental cause that she also pursues and which benefits many more people. She does not do so necessarily because she trades off the value of being with her children against those values she would produce for other people, nor because she believes that her action produces a more valuable state of affairs. It is rather that her obligations to her children in some cases over-ride her other obligations. In other contexts, for example, where direct action to stop the release of toxic waste in a nearby community means a risk of jail and loss of time with her family, she may reasonably decide that the matter is so important that it justifies the risk. She makes a judgement about which obligations have greater seriousness or importance or which have greater stringency. There is a sense in which one can say that an agent compares the importance of different obligations that make demands on her. However, the idea of trading off values in order to arrive at the outcome with the highest value does not describe the kind of comparisons that are being made. Rather it will be a matter of the relations in which one stands with respect to different individuals and groups and considering the competing claims they make on you as an agent. General claims of need will matter, but so also will particular relations to individuals with whom one has special ties.

Consider conflicts from a virtues perspective. From a virtues perspective what matters is what kind of person one should be. So, for example, take the choice offered to us by the highwayman, or armed robber, who declares 'your money or your life'. (Clearly these highwaymen are schooled in economic theory and have signed up to the trade-off model.) Suppose a person refuses, and is shot. Is it really plausible to say that she values her money above her life? Not her money, you will hasten to say, but her dignity or some such thing. Even so, the idea that this is a form of exchange does not come easily. Rather, giving way in a situation such as this is simply unthinkable, not compatible with the woman's perception of the kind of person she is. So she 'sacrifices' her life. And sacrifice is precisely not a trade-off, but rather a refusal to engage in trade-offs. As Steven Lukes observes:

> Trade-off suggests that we compute the value of the alternative goods on whatever scale is at hand, whether cardinal or ordinal, precise or rough and ready. Sacrifice suggests precisely that we abstain from doing so. Devotion to the one exacts an uncalculated loss of the other.
>
> (Lukes 1997: 188)

Once we begin to engage in trade-offs we have lost the goods in question. A person may refuse to even begin to negotiate over her dignity. If she were to do so it would already be lost. Likewise, with some of the constitutive incommensurabilities we discussed earlier. A person for whom relations to friends and kin really mattered would refuse to begin to negotiate how much she would be willing to accept for betraying them. To do so would already be to abandon the central commitments of her life. Similar points apply at the level of community. It is a common observation that from a strict cost-benefit perspective there is something odd about a local community spending resources and incurring the risk to rescuers' lives to save trapped miners, or a few people in distress in a storm. However, there is no question of acting here in terms of some trade-off schedule. The very thought is mistaken. We act in accordance with what expresses our values of solidarity. This is not to say that anything goes. If again the risks are extreme we may decide to call off the rescue attempt – what is courageous becomes foolhardy. However, it is our self-understandings about the kinds of individuals and communities we belong to that do the work here.

Structural pluralism

In the last section we contrasted consequentialist, deontological and virtues perspectives on resolving conflicts. We noted that each of these perspectives can take pluralist forms. A consequentialist can assume many values in determining what makes for a good state of affairs. A deontologist can assume many basic obligations irreducible to each other. A virtues ethicist can assume that there exist an irreducible plurality of virtues. However, as these theories are usually presented in textbooks, each assumes that there is a structure to ethical theory with certain ethical primitives. They are reductionist in that they offer different accounts of what these primitive concepts of ethics are, and then attempt to show how other ethical concepts can be reduced to those primitive concepts. As we noted in chapter 3, the primitives can be expressed in term of the question each takes to be basic in ethics, and the answer each provides:

Consequentialism: 'What state of affairs ought I to bring about?' The primitives of ethical theory are states of affairs. What is intrinsically good or bad are states of affairs. Actions and states of character are instrumentally valuable as a means to producing the best state of affairs.

Deontology: 'What acts am I obliged to perform or not perform?' The primitives of ethical theory are the acts of agents. What is intrinsically good are certain acts we are obliged to perform and what is intrinsically bad are certain acts which are impermissible. States of character are instrumentally valuable as dispositions to perform right acts. A state of affairs is right if it is the outcome of morally just acts.

Virtues ethic: 'What kind of person should I be?' The primitives of ethical theory are dispositions of character. The basic good of ethical life is the development of a certain character. A right action is the act a virtuous agent would perform; the best state of affairs is one that a good agent would aim to bring about.

Why should one believe that there are primitives in ethics of this kind? One pull towards the search for such primitives is a certain ideal of what an ethical theory should be like. Like a scientific theory it should have a particular structure. It should consist of a number of basic explanatory concepts, and then show that others can be derived from those basic concepts. The structure of the theory will give some kind of order and organisation to our ethical reflections on particular cases.

The need to bring some kind of order to our ethical reflection is a real one. It is true that in ethical choices we find ourselves faced with conflicting considerations and there is a need to bring some kind of organisation to the different considerations that gives each its due. However, it is far from clear that such order must be supplied by the ethical equivalent of a scientific theory complete with theoretical primitives. There are at least two sets of reasons for being sceptical. First, the different ethical primitives cannot be logically isolated in the way that such theories require. Ancient virtue ethics did not aim to offer a reductionist theory of this kind (Annas 1993: 7–10). One reason is that the specification of ethical virtues requires reference to other goods and harms (O'Neill 1997a). One cannot state why a virtue like courage is a virtue without mentioning that it involves standing firm against certain independently defined harmful states of affairs. However, the same is true of other putative ethical primitives. One cannot say what is morally wrong with acts like torture without reference to the pain from which it issues. Some states of affairs themselves can only be characterised as wrong in terms of their involving failure to respect persons, or cruelty of character. The idea that one could get one set of ethical primitives to do all the work is implausible. Second, attempts at reduction involve the elimination of dimensions of value, or at least failure to give them their due. As a result they are false to the nature of the moral conflicts we find ourselves in. For example, there can be situations in which hitherto virtuous people find themselves faced with doing terrible acts to save the good – the problems of 'dirty hands' are of this kind. To save one's moral character would be mere squeamishness. However, admirable people can act with integrity in contexts where to do so is hopeless and leads to no independently better state of affairs. Consider for example those who battle against insuperable odds for environmental justice, or to protect some marginal community – say the women prepared to be flooded in Narmada. Even where we know they will fail we can admire them.

If we deny that there are ethical primitives which can structure ethical theory we will be committed to another form of pluralism over and above that which we have already considered. One might call it structural pluralism. According to this view, pluralism can exist not just within different dimensions of value – between different valued states of affairs, between different obligations, between different virtues – but also between them. (Light (2003) previously termed such a view 'metatheoretical pluralism'.) What this means is that there can be situations in which consequentialist, deontological and virtues-based considerations, respectively, can themselves point us in different and indeed conflicting directions. If we defend such a form of structural pluralism, this raises with increased urgency the question: what role do we give to ethical reflection? If this is the reality of our moral predicament, how are we to think about reasoning ethical problems through? Well one answer is clearly ruled out. It is not a matter of coming up with some theory akin to a scientific theory, a set of primitives from which we can infer what action is to be performed. However, as we shall argue, the absence of a theory of that kind does not rule out the possibility of reasoned reflection which gives different considerations their due. Neither does the absence of a measure of value rule out the reasoned accommodation of conflicting values. Let us consider next then how ethical reflection under such conditions might proceed.

Choice without commensurability

If we reject the idea of resolving conflicts through the use of a single measure of value that allows us to trade off values, or the use of a general umpiring principle that orders values, how are we to resolve conflicts? How should we make rational choices?

Consider our two objections to the metaphor of trading-off values. The first concerned the ways that some values may not be convertible into others. In particular there are values we cannot express in monetary terms. To do so would be to betray those values. Some social relations and value commitments are constituted by a refusal to trade. To make this claim is, *inter alia*, to make a point about what is and is not rational: it is to note that there are more or less rational ways of expressing one's values. Actions can be rational and irrational in the ways in which they express values. Expressive accounts characterise actions as rational where they satisfactorily express rational evaluations of objects and persons: 'Practical reason demands that one's actions adequately express one's rational attitudes towards the people and things one cares about' (Anderson 1993: 18). The point here is that actions are not just instrumental means to an end, but a way of expressing attitudes to people and things. Hence some of the problems with monetary measures noted earlier. For example, one cannot express one's concern for one's children by saying, 'I love them so much that it would take at least one million pounds to give them up.' That would be an expression of potential betrayal of your relation to them, not an expression of love. Likewise with many values – to put a price on them is to express where one would be willing to betray them.

Our second objection to the trading-off position is that we do not need to make choices by measuring losses and gains on various values according to some measure, and then deciding which produces the highest total value. Our choices are not a matter of mathematics in that sense. Rather, choices can be a matter of considering competing claims that individuals in different relations make on us. This picture of competing claims can be generalised. The great promise of utilitarianism is that it appears to offer the means of making a choice through moral mathematics. But that is not how we make most choices. We are faced rather with a number of reasons or grounds for conflicting options, and we have to judge which reasons count most strongly in that choice. We will listen to others because they bring to our attention claims and considerations we have overlooked. We listen to the arguments and make a judgement.

Consider again the example of forest management we discussed earlier in the chapter. We noted that in such cases individuals representing different concerns and interests argue their corner, and the decision emerges from such arguments. Now, as we also noted, there may be a number of problems with such argumentative procedures in practice. In particular, the power and influence of different groups may be unequal. However, the idea that the best way of arriving at an answer is to argue it out is not as such objectionable. Indeed what is wrong with the inequalities of power is precisely that they get in the way of our reaching decisions by rational argument. It is through deliberation rather than measurement that we make such decisions. This is not to say that measures have no role. Some considerations will call on measures – say of species loss, or of poverty in local communities. Others will not: for example, the aesthetic properties of landscapes do not easily lend themselves to measurement. Neither do the meanings and social memories a place might have. It is not the practice of quantification or measurement as such that we are criticising here, but rather a vision of social choice as one that ideally proceeds through calculation rather than deliberation.

The appeal to deliberation through argument calls on what is sometimes described as a 'procedural' account of rational choice. Procedural accounts of practical reason take an action to be rational if it is an outcome of rational procedures: 'Behaviour is procedurally rational when it is the outcome of appropriate deliberation' (Simon 1979: 68). Rational behaviour is that which emerges from deliberation that meets the norms of rational discussion. Given a procedural account of rationality, what matters is the development of deliberative institutions that allow citizens to form preferences through reasoned dialogue, not the refinement of ways of measuring given preferences and aggregating them to arrive at a putative 'optimal' outcome.

What can we expect from a theory of rational choice?

There is a widespread assumption that reason can and should determine a unique best optimal solution in practical affairs. But that is simply to make unreasonable demands on reason. It is not a demand that we make upon reason when we make choices between different theories. We recognise that rational argument and existing evidence sometimes underdetermine choices in scientific beliefs. Hence rational individuals can make different judgements on available evidence. While there is a difference between theoretical and practical reason – the former ends in beliefs that are reversible, the latter in an act which may be irreversible – there is no reason to require of practical reason something that we do not require of its theoretical counterpart. Like theoretical reason, practical reason might sometimes determine a unique optimal solution, but equally it might not. A distinction needs to be drawn here between complete and partial resolution. By complete resolution we mean a resolution through which, by the use of judgement in a particular context, we can arrive at a unique best choice. By a partial resolution we mean a resolution that arrives not at a unique answer, but at a set of admissible solutions which themselves are not ordered. It is possible that one might simply have a variety of options, each with their own bundle of goods, each coherent and making sense, and with no ordering between them. Consider, for example, someone facing a version of the old choice between a life of contemplation and a life of action. They are deciding, say, between going to university and realising their not insignificant mathematical abilities or signing with a major sports team and developing their considerable footballing talents. Now, there are a series of comparisons that might be made here, but in the end there may simply be no 'best' choice. They are simply different realisations of a good life. Each sacrifices goods. One may regret, with good reason, the losses one incurs through the choice one makes. One is not indifferent between the options – it matters a lot which is chosen and a person is likely to agonise. But in the end one cannot order them. Often, admissibility is all that is possible and this is one source of the possibility of plurality in good lives that can reasonably be chosen. This is true also of social goods. This plurality of admissible solutions can be constrained but not eliminated by procedural and expressive accounts of rational choice. There are different lives that adequately express different bundles of goods and that have survived full reflection. Given a plurality of intrinsic goods, it is possible that different life plans, ways of life, and cultures, arrive at quite different bundles of goods which are still coherent, admissible and admirable (O'Neill 1995; Light 2003). None of this need be a source of worry about practical reason. Rather it points to a reasonableness about its limits in determining a specific outcome.

To say this is to recognise that reason cannot ensure for us in advance that conflicts of values can be resolved and tragic choice avoided. The question 'How do we resolve moral conflicts?' may be the wrong question. We may at the end of reflection have to accept that we are in a situation in which whatever we do a wrong will be done. There is no reason why reason should make ethical and political reflection easy. It may be more a matter of learning to live with ethical conflict rather than resolving it. And living with it may be uncomfortable. It may result in a number of ethical residues – shame at having to do something that runs against our deepest conception of the kind of person we aspired to be, or regret that we act with integrity in a situation in which we know the worst will befall us, or guilt where we find ourselves wronging some individual.

In contrast, the trade-off model fails utterly to explain the toughness of tough decisions. Indeed, such models serve only to conceal and suppress the toughness of choice. The point can best be registered by considering the typical 'fall-out' from a tough decision – namely anguish. Suppose that tough decisions are indeed trade-offs. As previously mentioned, this certainly means that something perceived as desirable has been given up or forgone. And no doubt this gives us cause to regret what we have had to give up. But at any rate the exchange has been made, and we have got the best deal. There are hardly grounds here for anguish over the decision itself. Yet anguish is precisely what one might expect in the wake of a truly tough decision. This may stem in part from the 'residue' of tough decisions – perhaps a perception that whatever we do would be wrong; but it also stems in part precisely from the *absence* of a yardstick – a circumstance that leaves us lost and confused, can induce trauma, and can even break our spirit (see, further, Holland 2002).

For those sensitive about the need to develop our understanding of ethics and morality so as to encompass the value of nature, an awareness of conflict and a sense of loss may already be acute. The dominance of a particular style of utilitarian thinking in environmental policy, principally through cost-benefit analysis, has disguised rather than revealed the toughness of many of the decisions we have had to make. By reducing values to a matter of the satisfaction of preferences whose strength can be measured in monetary terms, it has reduced environmental tragedies such as species loss and the degradation of place and the means to basic human livelihood in many parts of the world to a matter on a par with the removal of items on a supermarket shelf for consumer satisfaction. A sense of the enormity of the many problems we are facing has been lost. At the same time the approach has concealed some of the difficult choices that can exist in the face of demands to realise the aspirations of humans to safe, secure and flourishing lives, and to sustain the richness of life in the natural world. We noted in the opening chapter of this book that we live from, in and with the environments in which we find ourselves. The competing demands that these different relations place upon us leave us with hard ethical and social dilemmas. There are no algorithms to follow of the kind expert decision procedures often promise that can ease the difficulty of such dilemmas.

We have no easy solutions to these dilemmas, but we have concluded from these last five chapters that the predominant utilitarian framework for attempting to address them is not the right framework. In the next part of the book we examine the main alternative in recent environmental philosophy to this utilitarian framework which suggests that our environmental problems require a radically new ethical theory that breaks with the 'anthropocentrism' of traditional Western philosophy. We will suggest that this account fails and does so in part in virtue of sharing the assumptions about the nature of ethical theory we have criticised in this chapter. In developing this criticism we attempt to begin the process of building an account of how to better understand, evaluate and accommodate competing environmental values. In line with our conclusions in this part of the discussion, our answer will lie in a proper appreciation of the complexity of values, rather than an attempt to simplify them (cf McNaughton, 1988: 130). While agonistic in character, we think it will prove, in the end, more fruitful in grounding a more comprehensive framework for better environmental decision making which we begin to outline in part three.

 # PART TWO
A new environmental ethic?

The moral considerability of the non-human world

New ethics for old?

In part one of this book we have considered the broadly utilitarian approaches to decision making that are used to justify the main policy-making instruments employed in public decisions about the environment, such as cost-benefit analysis. In chapter 3 we examined the constraints on a policy of maximising total welfare that have been articulated, in different ways, from within deontological and virtue ethics. In chapter 4 we argued that a purely maximising approach cannot capture the distributive dimensions of good environmental policy. In chapter 5 we argued that the dominant utilitarian approaches to environmental policy are not consistent with the existence of plural and incommensurable values. We also gave reasons for scepticism about the very idea that reasoned ethical reflection should take the form of providing the ethical equivalent of a scientific theory, complete with theoretical primitives, from which our specific obligations could be deduced.

In the second part of this book we will turn to the mainstream alternatives to these utilitarian approaches that have been developed within recent environmental ethics. The central move of most work in environmental ethics has been to argue that there is a need for a 'new environmental ethic' that breaks radically with what are taken to be the existing Western traditions of ethical theory. (For surveys of this work, see Ouderkirk 1998; Wenz 2001; Light 2002a and Palmer 2003). Early on, environmental ethicists such as Richard Sylvan (then Routley) urged that an ethic for the environment needed to come up with an entirely new approach to assessing value in the world, since the overwhelming 'anthropocentrism' of traditional ethical theories had in part contributed to the growing environmental crisis by creating theories that were incapable of formulating moral reasons for protecting nature. Here is an opening volley from a classic early paper in the field:

> It is increasingly said that civilization, Western civilization at least, stands in need of a new ethic (and derivatively of a new economics) setting out people's relations to the natural environment, in [Aldo] Leopold's words,

'an ethic dealing with man's relation to land and to the animals and plants which grow upon it.' It is not of course that old and prevailing ethics do not deal with man's relation to nature; they do, and on the prevailing view man is free to deal with nature as he pleases, i.e. his relations with nature, insofar at least as they do not affect others, are not subject to moral censure.

(Sylvan 1973: 47)

Such views were enormously influential and did much to shape what counts today as environmental ethics as a distinctive subfield of philosophy (cf. also Routley, R. and V. 1980).

While the view that we need a 'new environmental ethic' claims to break with the western tradition of ethical philosophy, there is one assumption that it does share with its immediate predecessors in that tradition. It shares the same view of what an ethical theory should look like. Consider for example Sylvan's characterisation of what an ethical theory should look like:

An ethical system S is, near enough, a propositional system (i.e. a structured set of propositions) or theory which includes (like individuals of a theory) a set of values and (like postulates of a theory) a set of general evaluative judgements concerning conduct, typically of what is obligatory, permissible and wrong, of what are rights, what is valued, and so forth. A general lawlike proposition of a system is a principle; and certainly if systems S_1 and S_2 contain different principles, then they are different systems.

(Sylvan 1973: 47)

Sylvan expresses with some clarity here the view of what an ethical theory should look like that we criticised in the previous chapter. Like a scientific theory – or in Sylvan's case a logical theory (Sylvan's other major interest was in relevance logic) – an ethical theory should consist in a set of basic postulates or principles from which other lower-order ethical claims can be derived. To develop a new ethic is to develop a new theory that starts from different primitive postulates. The two primitive postulates that are taken to define traditional anthropocentric approaches to ethics are those that concern (i) the class of beings who deserve moral consideration and (ii) the domain of beings or states of affairs that have intrinsic value. Accordingly the new non-anthropocentric ethic has been built around two claims: first that the class of beings to whom moral consideration is owed needs to be extended beyond human persons; second that nature has intrinsic value.

In the next three chapters we shall examine the claim that we need a new environmental ethic. In this chapter we consider arguments for the claim that we need to extend the class of beings to whom moral consideration is owed. In chapter 7 we consider the claim that nature has intrinsic value, and the various meta-ethical arguments this has raised. In chapter 8 we examine what sense might be made of the claim that a being or state of affairs has value in virtue of being natural. As may be evident already, we propose to challenge the claim that our environmental crisis demands a new uniquely environmental ethic. We will suggest in this chapter and the next that its plausibility stems from the picture of ethical theory that it shares with some of its philosophical opponents. If one rejects this view in favour of the pluralist perspective we outlined in the last chapter, then much of the case for a new environmental ethic disappears. Our environmental crisis might require fundamentally different political and economic institutions. It may also require some changes in the ethical perspectives we bring to our dealings with both fellow humans and the wider non-human world. However, the idea that this requires some new ethical *theory* is mistaken. The ethical perspective we need to start from is not one that invokes radically new foundational postulates, but the human scale of values that we bring to our everyday encounters with human and non-human beings and environments with and in which we live. We need to start with the kinds of relations to our environments we outlined in the first chapter of this book, and the normative vocabularies we use in expressing those relations.

Moral considerability

The debates around the scope of moral considerability have, since a classic paper by Kenneth Goodpaster (1978), been concerned with how to complete the following statement:

> For all x and for all y, x is owed moral consideration by y if and only if x is F and y is G.

In this statement what is entered under G will specify the conditions for moral agency, that is, the properties a being must possess to be a moral agent. What is entered under F will specify the conditions for moral considerability, that is, the properties a being must possess to be an object of moral consideration. There is a long tradition of debate in moral philosophy around the conditions of moral agency. Most views will accept that to be a moral agent one must have capacities for rational deliberation and choice. Some will add one or more further conditions: capacities for particular emotional responses to others such as sympathy or empathy; capacities to form second-order desires about one's first order desires; capacities constitutive of freedom of the will; capacities to recognise oneself as a moral agent. How ever those debates run, the conditions for moral agency need to be distinguished, at least analytically, from the conditions of moral considerability – the properties a being must have to be a direct object of moral consideration. There is no reason to assume in advance that only moral agents should be deserving of moral consideration. One standard way of making this point is to say that it is not the case that only moral agents should be moral patients. In many standard ethical accounts, for example, we do not say that we owe no moral obligations to infants or the mentally impaired because they lack capacities for reason found among adult humans. Because such beings are moral patients they are owed moral consideration regardless of the probability that they will grow up (in the case of infants) or recover from some impairment. There are of course arguments at the margins: Do we owe moral obligations to human foetuses? Such worries are at the core of many philosophical and public debates about private morality and public policy. But what is not at issue in such debates is the question of who will inherit responsibilities for such entities if they are found to be morally considerable. Similarly, the central debates in environmental ethics have not concerned the conditions of moral agency but those of moral considerability. What is the class of beings to whom moral consideration is owed?

While there is an analytical distinction to be drawn between moral agents and moral patients, as we noted briefly in chapter 2, one answer to that last question could still be that the class of beings to whom moral consideration is owed just is the class of moral agents. Thus, as we noted in chapter 3, on the Kantian account the class of beings that have moral standing is that of rational persons:

> For all x, x is deserving of moral consideration from a moral agent if and only if x is a rational person.

This much is implied by the passage we quoted earlier:

> Beings whose existence depends, not on our will, but on nature, have none the less, if they are non-rational beings, only a relative value as means and are consequently called *things*. Rational beings, on the other hand, are *called* persons because their nature already marks them out as ends in themselves – that is, as something that ought not to be used merely as a means – and consequently imposes to that extent a limit on all arbitrary treatment of them (and is an object of reverence).
>
> (Kant 1956: 2. 90–91)

What marks out rational persons as beings to whom moral consideration is owed is their capacity to make rational choices about their own lives, their autonomy. This capacity confers a dignity on persons. They are ends in themselves, beings who have moral standing.

The basis of Kant's argument for extending moral consideration to all rational agents is that this is a requirement of rationality. In the first place, rationality requires consistency. So, if a rational agent acts on a given principle on some given occasion, then that agent should act on that principle on any other occasion of the same kind. Thus principles must be generalisable. But the reasons afforded by principles are also impersonal. If a principle gives me a good reason to act on some occasion it gives a good reason to anyone else in the same situation. Therefore I should act only on those principles I can universalise, that is, make into a universal law that applies to everyone. Hence Kant's formulation of what is known as the categorical imperative:

> 'Act only according to that maxim by which you can at the same time will that it should become a universal law.'

Using this basic principle Kant argues, further, that all rational agents are committed to treating other rational agents as ends in themselves. His reasoning runs as follows. Every rational agent is necessarily committed to treating himself or herself as a rational agent who governs themselves by their own reason, that is, as an autonomous being who is an end in themselves. So since reasons are impersonal, it follows that every rational agent will recognise that all other rational agents are also committed to treating themselves as ends in themselves. Hence, all rational agents are committed to treating other rational agents as ends in themselves. Thus Kant offers another version of the categorical imperative:

> 'Act so that you treat humanity, whether in your own person or in that of another, always as an end and never as a means only.'

We will not consider in detail here whether this basic version of the argument works – although the reader might reflect on whether the fact that all rational agents are committed to treating themselves as ends in themselves does entail that they are committed to treating each other as ends in themselves. While many in the environmental ethics debate, unlike Kant, have wanted to extend moral considerability beyond the class of rational agents, many have also appealed, as we shall see, to the basic structure of argument that Kant employs to get to that conclusion. Where they have differed from Kant is in the premises that they employ.

The main source of criticism of Kant in the environmental ethics literature has been to the corollary that Kant takes to follow from his position about where the boundary of moral considerability stops. Non-rational beings, including non-rational animals and living things, are, in contrast to rational persons, mere things that have only instrumental value for persons. They are not ends in themselves and persons have no direct duties towards them. The contrast between rational human agents and other creatures is drawn thus in Kant's lectures on anthropology:

> The fact that the human being can have the representation 'I' raises him infinitely above all the other beings on earth. By this he is a person . . . that is, a being altogether different in rank and dignity from things, such as irrational animals, with which one may deal and dispose at one's discretion.

(Kant 1968: 7. 127)

Non-rational beings lack the capacity for self-consciousness and, in virtue of that fact, do not have direct moral standing. They only have value as a means to an end, not as ends in themselves. Moral agents have no duties towards them. What might appear to be duties towards them are in reality indirect duties towards our fellow humans:

> If a man shoots his dog because the animal is no longer capable of service, he does not fail in his duty to the dog, for the dog cannot judge, but his act is inhuman and damages in himself that humanity which it is his duty to show towards mankind. If he is not to stifle his human feelings, he must practise kindness towards animals, for he who is cruel to animals becomes hard also in his dealings with men.
>
> (Kant 1979: 240)

This passage is open to different interpretations. The last sentence suggests a simple empirical claim that those who are cruel to animals are more likely to be cruel to human beings and that cruelty is wrong in virtue of those empirical consequences. However, the previous sentence suggests a more subtle position, that a person who acts with cruelty to an animal fails to show or develop the moral character that is proper to a human agent; the person damages his moral character. The last sentence then points to the implications of that failing in moral character. Whichever account one assumes, the central argument appears to have problems. Why might not the act of shooting the dog, or inflicting pain on the animal, be an exercise in catharsis that would render a person better disposed to other human beings? The obvious answer is that the intentional infliction of pain on the non-rational animal involves some wrong such that the nature of the act itself damages the moral character of the person who engages in it. But why should that be the case if the act in itself involves no direct wrong to another creature? Indeed why is the normative term 'cruelty' an apt description at all? It is precisely because it is an act of cruelty that wrongs another being – and is objectionable as such – that it reveals a failure of the moral dispositions of the agent.

Whereas Kant's ethics restricts the scope of moral considerability to rational persons, as we noted in chapter 2, classical utilitarianism extends the constituency of moral considerability beyond rational persons to include any sentient being. It is the capacity to feel pain and pleasure that marks the conditions of moral considerability. As Bentham famously put it in the passage we quoted in chapter 2, 'the question is not, Can they reason? nor, Can they talk? but, Can they suffer?' (Bentham 1789: 17. 283). Thus the utilitarian version of moral considerability runs as follows.

> For all x, x is deserving of moral consideration from a moral agent if and only if x is a sentient being.

For the classical utilitarian pleasure is good and pain is bad. It makes no difference whose pleasure or whose pain it is. Therefore the utilitarian is committed to impartiality. Anyone who may be affected by an action is to be considered on equal terms with any other – each is to count for one and none for more than one. Distance in time or place makes no essential difference. Thus geographically remote people must be considered; partiality to members of one's own nation or ethnic group is ruled out. Temporally remote people – future generations – must be considered; partiality to ones contemporaries is ruled out. Likewise species membership is irrelevant. All beings that are capable of suffering and enjoyment count, and not just human beings.

As we remarked earlier the most notable modern proponent of utilitarianism who has developed this position with respect to non-human animals is Peter Singer. Why should moral consideration be extended to sentient beings and no further? Singer offers the following commentary on the passage from Bentham:

> The capacity for suffering – or more strictly, for suffering and/or enjoyment or happiness – is not just another characteristic like the capacity for language, or for higher mathematics. Bentham is not saying that those who try to mark 'the insuperable line' that determines whether the interests of a being should be considered happen to have selected the wrong characteristic. The capacity for suffering and enjoying things is a prerequisite for having interests at all, a condition that must be satisfied before we can speak of interests in any meaningful way. It would be nonsense to say that it was not in the interests of a stone to be kicked along the road by a schoolboy. A stone does not have interests because it cannot suffer. Nothing that we can do to it could possibly make any difference to its welfare. A mouse, on the other hand, does have an interest in not being tormented, because it will suffer if it is.
>
> (Singer 1986: 221–222)

A point to note about this passage is that Singer takes sentience to mark the boundary of moral considerability on the basis of a yet more basic assumption: that a being deserves moral consideration if and only if it can meaningfully be said to have interests of its own. His argument, then, starts from the following more basic assumption:

> For all x, x is deserving of moral consideration from a moral agent if and only if x is a being that has interests of its own.

Why employ this principle of interests? The answer is that morality demands impartiality. A moral agent is one who gives equal consideration to the interests of all affected by some action. Hence a being falls under the scope of moral consideration if and only if that being has some interests to be considered. Inanimate objects like stones have no interests. Hence, they are not morally considerable. More significantly in the context of environmental reasoning, Singer also assumes that non-sentient living things do not have interests in the sense that counts for ethics. A being can have interests if and only if it is capable of suffering and enjoyment. Hence all and only such sentient beings are morally considerable. Again, as with Kant's argument, while later theorists in mainstream environmental ethics have criticised the assumptions that Singer has made in confining moral considerability to sentient beings, the general form of Singer's argument has remained influential.

As we noted in chapter 3, the conflict between utilitarian and Kantian ethics is not just about the scope of moral considerability. It is also about the nature of moral standing that is accorded to those who are taken to have moral considerability. A central feature of Kant's account of moral standing which has been a source of its lasting influence in debates with utilitarianism is that it blocks the treatment of beings as a means to some greater total good such as increasing total welfare. Rational persons are ends in themselves in the sense of having the capacity to govern themselves according to their own reason. In virtue of this capacity they have a dignity. Respect for that dignity rules out treating them merely as a means either to the good of others or indeed their own good. This moral standing is often expressed in terms of rights. Individuals have rights that cannot be overridden for the greater welfare. One of the continuing attractions of the Kantian approach in environmental ethics is that, while Kant's account of who has moral standing is much narrower than that offered by the utilitarian, the form of moral standing it offers promises to be much stronger. Hence while it looks unpromising at first, if a Kantian account of moral standing can be extended to non-humans it offers much stricter forms of protection to those beings. Consider for example the use of animals in medical research. From a utilitarian perspective, if the total welfare gained is greater than the welfare losses involved in the animal experiments then there is nothing in principle wrong in those experiments. The only proviso is that one must not be speciesist: in other words, if the animal experiment is justified, so also is the same experiment on a relevantly similar human being (Singer 1986: 223–225). A rights perspective on the contrary will argue that if a being has rights then it is wrong to inflict intentional suffering on that being even if it leads to greater total welfare. It is thoughts of this kind that are the motivation for Regan's attempts, which we discussed in chapter 3, to extend moral standing in the Kantian sense beyond rational persons to any being who is the subject of a life.

Extending the boundaries of moral considerability

The central move in much recent environmental ethics has been to attempt to go beyond the circle of sentient beings drawn by utilitarians such as Singer and rights-based theorists like Regan and to extend the scope of moral considerability still further. Why might it be thought there is a need to thus extend the circle of considerability? At least part of the reason has to do with the implications for environmental policy of stopping at sentience. At a practical level nature conservation bodies and groups on the one hand, and animal welfare and animal rights activists on the other, can find themselves in conflict with each other. A central problem for nature conservation is non-endemic populations of animals that threaten local biodiversity. Consider for example the case of New Zealand which was one of the last places on the planet to be settled by humans. Prior to settlement it lacked both grazing and predatory mammals and consequently had a unique flora and fauna. The arrival of the first people in the tenth century saw the introduction of the pacific rat. European settlement has seen the introduction of a large number of non-endemic species – sheep, goats, possum, deer, horses, ferrets, stoats and weasels. Their introduction has had a large impact on native flora and fauna, and brought about the extinction of many species. On some estimations, for example, over 40 per cent of endemic bird fauna have been lost. The attempts to protect and restore the remaining endemic fauna and flora have often taken the form of large-scale culling programmes against introduced species that involve hunting, trapping and poisoning. Programmes like this are widespread across the globe where non-endemic species threaten local biodiversity.

Can the suffering caused to individual sentient beings by such programmes be justified? If one thinks that only sentient beings are morally considerable it is difficult to see how a justification is possible. For example, how can the protection of non-sentient plant and insect life justify the foreseeable suffering caused by the culling of sentient individuals? And how can the distribution of species or the protection of ecosystems matter as such, rather than just the total welfare of sentient beings, or the welfare of particular individual sentient beings? True, there are indirect consequences that a utilitarian might invoke. It may be that, in general, protecting biodiversity will lead to the well-being of sentient beings who depend on particular habitats. There is also the enjoyment that humans gain from the contemplation of species-rich habitats. Another case, made persuasively by Gary Varner (1998), is that certain kinds of animals ('obligatory management species') require active management in the form of 'therapeutic' hunting or else they will experience suffering from overpopulation (such as starvation) which far exceeds the pain that would be caused by culling them. However, while there might be some general rules of thumb here, in some cases such arguments have been characterised as instances of special pleading. The mass culls of mammals to protect indigenous fauna in New Zealand, for example, do not look justifiable on these grounds because they appear so indiscriminate. If one takes a perspective that starts from a more robust deontological form of animal rights, such culls look still more difficult to justify. From this perspective, individual sentient beings are taken to have rights that cannot be overridden for the general welfare. For these kinds of reasons it is not surprising that programmes to cull animals in the name of nature conservation and biodiversity are often opposed by animal welfare and animal rights groups as unethical even where some philosophical proponents of these theories, such as Varner, may find them justifiable or even obligatory.

In the face of such arguments one response by philosophers who want to defend the practices of nature conservation bodies has been to suggest that we need to extend the scope of moral considerability. There are two distinct moves that might be made here. The first is to extend the scope of moral considerability from sentient beings to include all individual living things. Robin Attfield (1987) attempts to make that move from within a consequentialist perspective and Paul Taylor (1986) from within a broadly-speaking deontological perspective. Both positions still make individuals the object of moral consideration. Where they differ is in the account of moral standing they extend to those individuals. The second move is to argue that the locus of moral appraisal should not be solely on individuals. There are two versions of this thought. One version is that collective entities such as species or ecosystems should themselves be direct objects of moral consideration. A second is that the good of a community is in some way involved in the adjudication of conflicts of interest that arise between different individuals who form parts of a collective entity. Both versions share two common intuitions. First, if environmental ethics is to be informed by the science of ecology then it must be recognised that the possible normative implications of the science of ecology rarely, if ever, concern the welfare of every individual thing in an ecosystem. As such, ecosystem management should not be aimed at preservation of the individual welfare of all things in an ecosystem (sentient or otherwise) and an ethics informing such management should also not be bound by such a restriction. Second, even a cursory glance at the policy imperatives of environmental advocates show that their principal concerns are over collective entities, such as species, ecosystems, and wilderness areas. The assumption here is that the most effective ethic which will inform our valuation of these entities will be one that gives us reason to find them directly considerable rather than only indirectly morally considerable given their role in the welfare of individual entities. The main source for this shift to some form of ethical holism has been Aldo Leopold's land ethic, which has been taken up by environmental ethicists such as J. Baird Callicott. We will discuss the moves further below.

We have argued that one central motivation for extending moral considerability beyond sentient beings has come from the need to provide an adequate account of conflicts within environmental policy. A second is a set of thought experiments or 'intuition pumps' that are taken to show that our understanding of who counts is wider than merely sentient beings. The most influential of these is Richard Sylvan's 'last man' thought experiments.

The *last man* example. The last man (or person) surviving the collapse of the world system lays about him, eliminating, as far as he can, every living thing, animal or plant (but painlessly if you like, as at the best abattoirs).

The *last people* example: The last man can be broadened to the last people example. We can assume that they know they are the last people, e.g. because they are aware that radiation effects have blocked any chance of reproduction . . . Let us assume the last people are very numerous. They humanely exterminate every wild animal and they eliminate the fish of the seas, they put all arable land under intensive cultivation and all remaining forests disappear in favour of quarries or plantations and so on.

(Sylvan 1973: 49–50)

Richard Sylvan claims that, if we agree that both the last man and the last people do wrong, then we have to reject what he calls 'human chauvinism' – the assumption that only humans count morally. One response might be to agree, but to argue that what is wrong is the harm done to other sentient beings. It is not difficult to see how both Singer and Regan, for example, could argue that there is a wrong on these grounds. However, the thought experiments might be adjusted to deal with such a response. Imagine that the same accident that stops the reproduction of the last people has the same effect on all sentient beings. The world humans leave will contain only primitive insects and plants. Would it be wrong to eliminate these as a final act? Or imagine the last man sits before the last great Oak tree and his final act is to watch it burn. Does he do wrong? If the answer to these questions is yes, that the acts of the last people and the last man are wrong, then who is wronged? It is not the last man, the last people, nor any other sentient being. Hence, the argument runs, it must be the case that at least some non-sentient beings also count morally; so we do need to extend the boundaries of moral considerability beyond sentient beings. The debates in mainstream environmental ethics have been largely concerned with different suggestions as to how that extension of moral consideration might be developed and justified.

One suggestion is to include non-sentient living individuals in the domain of moral considerability. Our statement on moral consideration should then read as follows:

For all x, x is deserving of moral consideration from a moral agent if and only if x is a living thing.

What reason might be given for extending the domain of moral consideration in this way? Well, one popular line of argument has been to employ the 'interests' account of moral consideration, but to suggest, further, that the domain of beings who have interests includes not just sentient but also non-sentient beings. The argument might be expressed as follows:

1. For all x, x deserves moral consideration from a moral agent if and only if x has its own interests.
2. A being has its own interests if and only if it is a living thing.

Hence,

3. For all x, x is deserving of moral consideration if and only if x is a living thing.

The crucial premise here – the one that justifies the extension of moral consideration beyond sentient beings – is the second premise, the claim that all and only living things have interests. On what grounds might all living things be thought to have interests? The answer is that while they do not have conscious desires or the capacity to feel and hence to suffer, they do have goods of their own that are independent of their instrumental value for the good of humans or other sentient beings. The thought here is broadly Aristotelian.

There is a sense in which we can talk of what it is for natural entities to flourish, and what is good and bad for them, without this being dependent upon human interests or those of other sentient beings (Attfield 1987; Taylor 1986; Rolston 1988: ch.3; Varner 1998). Thus consider the farmer faced with some weed that is bad for his livestock. He is concerned to eliminate the weed. Next consider uses of the phrase 'x is good for weeds' in this context. The term 'good for' can be understood in two distinct ways. It might refer to what is conducive to the destruction of weeds, as in 'this chemical spray is good for weeds'. The term 'good for' in this use describes what is instrumentally good for the farmer: given the farmer's interest in the flourishing of his livestock, the application of the spray destroys the weeds and serves his interests. Given that the spray is not harmful to his livestock one might also say that spraying the weeds is in the interests of his livestock quite independently of the farmer's own interests. It destroys a substance that is harmful to the animals and will cause them to suffer. The phrase 'x is good for weeds' in this context simply concerns what is ultimately in the interests of humans or sentient beings. So far we do not need to go beyond sentient beings in ascribing interests. However, the phrase can also be used in a second way. It can be used to describe what causes weeds to grow and flourish, as when our farmer, in answer to the question as to why there are so many weeds this year, replies in exasperation, 'mild wet winters are good for weeds'. This second use describes what is instrumentally good for the weeds, quite independently of the interests of the farmer or of his livestock. This instrumental goodness is possible in virtue of the fact that the plants in question are the sorts of things that can flourish or be injured (von Wright 1963: ch. 3). In consequence they have their own goods that are independent of human interests. What is the class of entities that can be said to possess such goods? In an influential passage von Wright identifies it with the class of living things: 'The question "What kinds or species of being have a good?" is therefore broadly identical with the question "What kinds or species of being have a life?"' (von Wright 1963: 50, cf. Taylor 1986: 60–71).

A living thing can be said to flourish if it develops those characteristics which are normal to the species to which it belongs in the normal conditions for that species. It can be said to have interests in the sense of having goods of its own. Non-living things have no such interests since there is no sense in which one can talk of their having their own good. They are not the sorts of things that can flourish or thrive. Significantly in the context of environmental ethics von Wright rejects the claim that social units like families and political associations have their own good in any literal sense. Their good is reducible to that of their members (von Wright 1963: 50–51).

The central move then by theorists like Attfield, Taylor, and Varner has been to take having its own good to be the crucial feature that marks out the class of beings who have moral standing: 'moral standing or considerability belongs to whatever has a good of its own' (Attfield 1987: 21). Thus:

> For all x, x deserves moral consideration from a moral agent if and only if x has a good of its own.

Taylor in this context talks of individual living things being 'teleological centres of life', in virtue of their having their own good (Taylor 1986). There are things that can benefit a living thing by allowing it to realise its good, and things that can harm it by thwarting the realisation of its good. We might therefore restate the argument above in terms of beings having a good of their own:

1. For all x, x deserves moral consideration from a moral agent if and only if x has its own goods.
2. A being has its own goods if and only if it is a living thing.

Hence,

3. For all X, X is deserving of moral consideration if and only if X is a living thing.

Attfield, Taylor, and Varner extend the class of moral considerability to include individual living things. None of them believes it should be extended further. Where they differ is in the account of moral consideration that they start from. We will summarise the position of the first two.

Attfield is a consequentialist. We should aim to maximise the flourishing of living things. However, his position does not assume a simple maximising sum for all living things – the interests of humans and other sentient beings would soon be swamped by those of other living beings. He modifies his account by introducing two further dimensions – psychological complexity and signifi-cance. He argues that the flourishing of psychologically complex beings is more important than that of less complex beings. However, in making decisions psychological complexity is not the only thing that matters. What also matters is the significance of the goods in question. A psychologically complex being is not justified in over-riding the significant interests of less complex beings in the pursuit of more trivial goods. So the pursuit of trivial human goods, say in tastes of food, cannot justify the destruction of those goods that are central to the flourishing of other living things.

Taylor's position on the other hand is deontological. The central thrust of his argument is to extend the respect that in Kantian ethics is owed to persons as rational agents to all living things as 'teleological centres of life'. Just as for Kant all persons are ends in themselves, so also for Taylor are all living things in virtue of having goods of their own and hence their own interests. They are owed the attitude of respect that is demanded of all beings who are ends in themselves. Again, like Attfield, he faces the problem of how to square that principle with allowing humans to live their lives without excessive self-sacrifice. He does so by employing a principle of self-defence to allow that human interests can over-ride the interests of other living things where significant human goods are at stake.

Both Attfield and Taylor are individualists. Those who count morally are individual living things. For reasons noted above, a number of philosophers have found this individualism problematic, again partly because it fails to capture some of the real issues in nature conservation. Many nature conservationists. for example, would want to argue that the preservation of some rare species of plant is of sufficient importance that it must sometimes be allowed to permit the destruction of other sentient and non-sentient beings. If individuals count simply as individuals it is not clear how that policy could be justified, or even entertained, since a rare species will necessarily comprise only a few individuals. One response has been to claim that we need to be able to refer to the goods of collective entities such as colonies, species, ecosystems, habitats and the like in ways that are not reducible to the goods of individual members. As we mentioned earlier, one writer who has been particularly influential in this regard was the philosophically inclined forester and ecologist, Aldo Leopold, whose land ethic makes the concept of a community central to ethics:

> All ethics so far rest on a single premise: that the individual is a member of a community of interdependent parts . . . The land ethic simply enlarges the boundaries of the community to include soils, waters, plants and animals, or collectively: the land.
>
> (Leopold 1949: 203–204)

The central ethical claim that has been taken up by later writers is the following:

> A thing is right when it tends to preserve the integrity, stability, and beauty of the biotic community. It is wrong when it tends otherwise.
>
> (Leopold 1949: 224–225)

Many have found this principle too vague, by itself, to be of much philosophical help. But whether Leopold was defending any full-fledged ethical theory, and in particular one that involves some form of holism about values, is a moot point. The idea that the land ethic did involve such a commitment was taken up, further explicated, and expanded on by later environmental philosophers. The biotic community has a good that is irreducible to that of the members. Leopold's reference to the health of the land might be taken to point in that direction: 'Health is the capacity of the land for self-renewal. Conservation is our effort to understand and preserve that capacity' (Leopold 1949: 221). It might be thought here that to talk of health in a literal sense entails that we can talk of the land being healthy or unhealthy in the same sense in which we talk of the health of the individuals that make it up. But while the health of the land depends in some way on the health of individuals that make it up, it is distinct from them and it might sometimes require the destruction of most individuals if self-renewal is to be possible. Thus, for example, it can be argued that forest fires might lead to the widespread devastation of flora and death and suffering for fauna, but be a condition for the renewal of the forest ecosystem as a whole. These kinds of argument can in turn lead to weaker and stronger forms of ethical holism. The stronger form would argue that collective entities are themselves directly objects of ethical consideration; an example, perhaps, is Rolston 1990. A weaker form (referred to as 'managerial holism') is that the good of a community is in some way involved in the adjudication of conflicts of interests between different individuals, but that it does not involve any claim about the moral standing of collective entities as such. Callicott appears to interpret Leopold's land ethic as entailing such a position:

> An environmental ethic which takes as its *summum bonum* the integrity, stability and beauty of the biotic community is not conferring moral standing on something *else* besides plants, animals, soils and waters. Rather, the former, the good of the community as a whole, serves as the standard for the assessment of the relative value and relative ordering of its constitutive parts and therefore provides a means of adjudicating the often mutually contradictory demands of the parts considered separately for *equal* consideration . . .
>
> (Callicott 1980: 324–325)

On this view the land ethic is taken to offer a standard through which the value of species diversity can be properly appraised. Still, since species are directly morally considerable on this view then it also involves a stronger form of holism. Both weak and strong forms of the claim have been the object of a great deal of debate in terms of their implications, again, for human beings in the natural world. One problem is that they appear to justify human diebacks for the sake of the '*summum bonum*' – the highest good of the biotic community. That implication was indeed embraced by some environmental philosophers, notably by Callicott himself in his earlier writings, although it is one from which he and others have now retreated. (On the debate that emerged from Callicott's early formulation see Callicott 1998; Jamieson 1998; Varner 1998.) Another worry involves the moral monism of some versions of holism, in particular Callicott's view, which is susceptible to the same sorts of criticisms we offered of individualist forms of monism in the last chapter.

How convincing are these attempts to extend moral considerability beyond sentient beings? There are a variety of different particular points that might be made. However, here we want to focus on some more general problems with the forms of argument involved. One set of problems concerns the basic argumentative strategy they employ. In particular, the various attempts at extension rely on different versions of the appeals to impartiality and consistency that are central to the Kantian and utilitarian traditions in ethics, and which we outlined above. To recap, the Kantian version of the argument runs like this. Since we regard ourselves as of ethical standing in virtue of the fact that we pursue our own good, then consistency demands that we extend ethical standing to any being that similarly pursues its own good. Since all living things have a good of their own, we must extend moral considerability to them. A similar appeal to consistency runs through utilitarian-based positions. Ethics demands that we give equal consideration to the interests of all affected by an action. Hence, we must extend moral standing to any being that has its own interests. Since all living things have their own good, they have interests. Hence, consistency demands that we must extend moral standing to all living things.

However, on closer inspection these appeals to consistency and impartiality look unconvincing. The problem here is that while it might be the case that having interests or having a good of one's own is a necessary condition for being morally considerable, it is not clearly a sufficient condition. There are some beings that we might recognise as having interests but not believe to be the kinds of being whose interests should be fostered. We can know what is good for X and what constitutes flourishing for X, and yet believe that X, under that description, is the sort of thing that ought not to exist and hence that the flourishing of X is just the sort of thing we ought to inhibit. The point is one made by Aristotle in his *Metaphysics* with respect to categories such as thieves, murderers and tyrants. One can state what it is to be a good tyrant, what it is for tyrants to flourish, and the conditions in which they will flourish, but believe tyrants, qua tyrants, have no claims upon us (Aristotle 1908: 1021b 15ff; cf. Rawls 1972: 402–404). That Y is a good of, or good for, X does not entail that Y should be realised, unless we have a prior reason for believing that X is the sort of thing whose good ought to be promoted. Thus just as, against hedonistic utilitarianism, there are pleasures that simply should not count, but rather should be the direct object of appraisal (e.g. those of the sadist), so one might think that there are interests we can quite properly refuse to count. One might think that there are some entities whose flourishing simply should not enter into any calculations, such as the flourishing of tyrants as tyrants. Correspondingly it is implausible to suggest that we human beings ascribe ethical standing to ourselves simply in virtue of having goods or interests of our own. While it might be the case that if a being has ethical standing then we have a prima facie obligation to promote its good, it is not the case that if a being has a good or interests then it has ethical standing. The fact that a being has a good, or interests, that might be promoted is what gives the recognition of ethical standing its point, but it does not logically ground that recognition. That a being has goods of its own might be a necessary condition for its having ethical standing. It does not follow that it is a sufficient condition. To make these points is not to assert that non-humans lack moral standing. However, if they do have standing, the claim that they have standing cannot be deduced from the fact that non-human beings have their own good. The farmer can accept that weeds have their own good. It is a separate question whether it is good for weeds to flourish.

A second set of problems concerns a basic conflict that arises in all these attempts to extend moral considerability. Stretching the domain of ethical considerability appears to result in a thinning of its content. To say that all living things count morally, just as humans do, appears to raise the status of living things. To say that all human beings count morally, just as all living things do, appears to lower the moral standing of human beings. All the theories we have discussed encounter some kind of problem when they are applied back to human beings, and they all rely on different adjustments to respond to those problems. Either a hierarchy of standing needs to be reintroduced or some strong self-defence principle invoked, if the defence of moral extension is not to have some quite problematic implications when applied back home. However, the difficulty here might again be traced back to the argumentative strategy involved. All the different accounts acknowledge that any ethical theory must allow us to treat different beings in different ways. Equality of consideration of interests does not entail equality of treatment. However, in extending moral considerability, the grounds for granting moral value to others do look to have thinned. The moral landscape seems to have been flattened. Indeed the very notion of moral consideration itself looks too thin to ground the very different type of response that is owed to different kinds of beings. From the point of view of virtue ethics, the focus purely on the possession of goods or interests appears to make all of ethics a matter of fostering just one virtue – beneficence – and avoiding one vice – malevolence. The variety of human virtues and vices, the different kinds of relations we have with the beings and environments we live with, seem all to disappear from view. In the next section we will suggest that the problem with the debate on environmental ethics has been with the very way that the question of moral considerability is posed. In the background are some assumptions about the nature of ethical theory that we criticised in the last chapter, and which we have seen in this chapter to inform the way the debate has been structured.

New theories for old?

We noted at the start of this chapter that defenders of the claim that we need a new environmental ethic have a particular picture of what that involves. What is required is a new ethical theory with new basic ethical postulates that will replace the primitive postulates of older ethical theories such as utilitarianism and Kantianism. In the last chapter we suggested that the search for ethical theories of this kind was a mistake. There is no reason to assume that rational ethical reflection should be modelled on the ideal of a scientific or logical theory with ethical primitives and basic theoretical postulates from which specific moral injunctions can be derived. This picture of ethics indeed is liable to distort and impoverish our moral language and responses. The problem with mainstream environmental ethics lies in what it shares with the ethical perspectives from which it takes itself to be escaping.

At the centre of this attempt at a new environmental ethics is the attempt to fill in a basic postulate about the domain of moral consideration.

For all x, x is owed moral consideration by a moral agent if and only if x is F.

Different theorists offer different accounts of how we should specify conditions for moral considerability under F. However, there is a prior question to be asked here about whether this is the right way to approach the question of our ethical relations to both the human and the non-human world. The approach assumes there is a single basic ethical predicate relation ' . . . is owed moral consideration by a moral agent' and that we have to specify some necessary and sufficient conditions for a being to fall under that relation. The basic motivating thought seems to be that we can come up with some set of necessary and sufficient conditions for being an object of generalised moral concern or consideration.

To see what is wrong with this approach it might be more fruitful to start from the other side, from the various appraisals we make of an actor in his dealings with the human and non-human worlds, with the virtue and vices he exhibits. Consider the last man and last people examples employed by Sylvan. An initial point to note about the example is that, like most ethical thought experiments of this kind, it is radically under-described. Real life conditions of uncertainty are absent. Consideration of what life would be like in the contexts described is erased. However, let us grant the conditions specified by the examples. The act itself still needs more specification. Is the last man's action an act of wanton destruction for the gratification of a final whim, an act of despair, an act of grief, or some kind of sacrament of last rites on the passing of life? We may not believe the act is right under any of these descriptions, but we will probably appraise it differently. Under the first description of the act we might say it displays the vices of insensitivity and wantonness, and exhibits a failure of the agent to appreciate the beauty and complex nature of living things. If animals are present we might add that it exhibits cruelty. Under the second description we would withdraw some of these claims but we might, perhaps harshly in the circumstances, refer to other failings, for example of steadfastness of character required to sustain hope and perception of what is valuable when things are going badly. Similarly with the other descriptions, we would apply different virtue and vice terms. There is a whole variety of terms that we would apply, each highlighting different excellences or deficiencies of character exhibited in the responses. In applying these virtue and vice terms we are not simply criticising the agent for failing to meet some list of moral excellences, for failing in the race to moral perfection as an end in itself. Indeed the very idea of morality as a race to ethical perfection is one that results in its own vices of insensitivity and self-absorption – of being concerned for others only as a means to the display of one's own virtue. It is the circumstances and objects themselves that evoke different responses. Steadfastness would not be a virtue if things did not go wrong and responses were not required to sustain what is valuable independent of the agent. Cruelty would not be a vice if suffering was not bad for those who are the object of cruelty. Virtue and vice terms invoke independent goods and ills. It is for this reason that we suggested in the last chapter that to see a virtues ethic as yet another kind of moral theory with different foundational postulates is a mistake.

An important point to note about our application of these virtue and vice terms is that objects evoke different proper responses under different descriptions. The concept of moral concern or moral consideration in the abstract does not capture the variety of responses that are required of different objects under different descriptions. For example, Kant is right that particular capacities for rational reflection about ends demand a particular kind of respect from others. However, it does not follow that all other objects are mere things that fall out of the domain of proper regard. It is rather the nature of that regard which changes. Likewise, sentient beings demand from us a particular set of relations of benevolence that non-sentient beings cannot evoke – one cannot be cruel or kind to a carrot. However, it does not follow that there are not other sets of attitudes and responses that are owed to non-sentient living things in virtue of their nature. As Aristotle notes, attitudes of wonder are owed to even the most humble of living things (Aristotle 1972, bk I, ch.5). Moreover there are more specific virtue and vice terms that are applied to persons under their more particular roles. Consider for example the specific virtues and vices of the gardener or the farmer – the care a good gardener shows in the ways she tends her vegetables and the soils on which they depend. Our plea here is to begin ethical reflection from the actual thick and plural ethical vocabularies which our everyday encounters with both human and non-human worlds evoke. If we start with a thick and plural ethical vocabulary we invoke a similarly thick and plural set of relations and responses appropriate to different kinds of beings. These are lost if we start from a picture of moral theory as an exercise in the derivation of specific moral norms from some set of moral primitive concepts or propositions.

The problem is revealed in the weakness of a certain kind of consistency argument that we discussed in the last section. There is nothing wrong with consistency arguments as such in ethics. However, those employed often appeal to a base set of descriptions that are too thin to do the job. Consider for example the idea that we can make 'having interests' the basis of moral concern for our fellow humans that we can then expand to all others. For reasons we have outlined at the end of the last section it cannot do the job. Having interests in itself is not a sufficient condition for the particular ethical response of concern that is required. Nor indeed is it the basis of our ethical concerns for our fellow human beings. For similar reasons, we will suggest in the next chapter that intrinsic value taken in abstraction also cannot do the work we require. These thin descriptions do not capture the nature of the different kinds of responses that we may owe to different kinds of beings.

We doubt there are any necessary and sufficient conditions for an object being an object of moral considerability until the particular kind of consideration is properly specified. It is however possible to specify the conditions a being must meet to be the object of a particular response. One can for example begin to specify the conditions a being must be able to meet to be an object of cruelty – a being must be sentient. Likewise, to be an object of paternalism a being must have capacities for autonomous choice. We might begin to specify similar conditions for at least some other responses. However, the idea that any such conditions can be specified for the generalised and underspecified concept of 'moral consideration' looks implausible. More significantly, any attempt to do so will privilege one particular kind of response at the expense of others. The result will be, as we suggested at the end of the last section, a flattening of the moral landscapes. Ethical reflection needs to start from the plurality of relations and moral responses that are owed to beings – not from some generalised and underspecified concept of 'moral consideration'.

The absence of some general moral theory of the kind promised by Kantian and utilitarian ethics will result in a much more difficult ethical life in which plural values pull us in different directions. As we noted at the end of the last chapter, rational reflection on ethical life cannot ensure for us in advance that conflicts of values will be resolved and tragic choice avoided. There may be no resolution and we may have to live with conflicts. The practical conflicts in conservation policy we have alluded to in this chapter may be of this kind. Philosophical reflection in ethics is not in the business of making ethical and political reflection easy. A proper appreciation of the complexity of values is to be preferred to simplifications that hide the wrongs we may be forced to do. Living with such conflicts may be uncomfortable and quite properly leave ethical residues. However, it will in the end offer a better starting point for understanding the difficulties of the decisions in question.

7 Environment, meta-ethics and intrinsic value

In the previous chapter we examined one of the central claims made by proponents of the need for a new environmental ethic. The claim is that a new ethic requires a new ethical theory that extends moral consideration beyond humans to include a variety of non-human entities. Debates of this nature, about who or what is entitled to moral consideration, form a part of what is often called 'normative ethics'. Normative ethics deals with first-order substantive questions in ethics, including those at the heart of environmental ethics which concern the significance of environmental changes, and the relative importance of those beings, human and non-human, who are affected. As we saw in part one of this book, normative ethics also typically involves an attempt to offer systematic theoretical frameworks for the justification and articulation of such claims. Consequentialist, deontological, and virtue theories are standard examples of such systematic theoretical frameworks. Proponents of the need for a new environmental ethic have tended to see themselves as engaged in a continuation of the very same project, but as offering different foundational postulates about the scope of moral considerability from those that have hitherto dominated mainstream ethics. In the last chapter we suggested some reasons for scepticism about the very nature of this project.

In this chapter we examine a second central claim made by those who advocate the need for a new environmental ethic, that is the claim that nature has intrinsic value. However, defenders of the claim that nature has intrinsic value have been engaged in arguments not just about the first order moral claims of normative ethics concerning who or what should count in ethical deliberations. They have also been engaged in second-order meta-ethical disputes about the nature and status of ethical claims. Meta-ethics is concerned with the nature and status of ethical claims. (For an overview of work on meta-ethics, see Darwall et al. 1992.) It does not deal with substantive questions in ethics but with questions about ethics – for example, whether or not ethical claims can be true or false, whether there is an ethical reality, and whether ethical claims are open to rational justification. For reasons we outline below, both defenders and critics of the claim that non-human nature has intrinsic value have assumed that this claim requires a form of realism about ethical claims, and hence the adoption of a particular meta-ethical stance. In this chapter we will examine the reasons for this shift from first-order questions within ethics to second-order questions about the nature and status of ethical claims. In doing so we will give some further reasons for being sceptical of the view that our environmental problems require a new ethic founded upon a revised metaphysical understanding of value. We will suggest again that a defensible approach to nature has to start from a human scale of values and from the rich normative vocabulary that has been bequeathed to us through our human engagements with the various environments we inhabit.

Meta-ethics and normative ethics

A central traditional question in meta-ethics is whether ethical utterances are assertions that can be true or false. The 'ethical realist' holds that ethical statements are descriptions of states of the world, and in virtue of being so they are, like other fact stating assertions, true or false independently of the beliefs of the speaker. On this view, it is the job of our ethical judgements to track properties in the world, to get something right about the way the world is. Against ethical realism stand a variety of views. One is the 'error theory', according to which ethical statements are indeed descriptions of the world, but they are all false; we project values on to a world and then talk as if they had independent existence (Mackie 1977). Another is 'expressivism', the view that ethical statements are not descriptions of the world at all; rather they serve to express the attitudes of the speaker towards the world. On this view if we sometimes say things like 'it is true that destroying rain forests is wrong', the phrase 'it is true that' serves only to give emphasis to the force of the attitude expressed. It should not be understood, as the realist supposes, as indicating that we are asserting something about states of the world that hold independently of the beliefs of the ethical agent. There are some ethical concepts that present an immediate problem for this expressivist view. We use certain specific ethical concepts like 'cruel', 'kind', 'cowardly', 'brave', to both describe and appraise states of the world. For example, to say that a farming practice is cruel is to make a claim of fact – that it involves the intentional infliction of pain. The expressivist answers that the descriptive component of specific concepts like 'cruel' can be prised apart from their evaluative component. We can analyse such concepts as the conjunction of a descriptive component that does the describing, and an evaluative component that expresses an attitude towards the act – a preference or feeling against it. To say an act is cruel is to say something like 'it causes intentional suffering and I disapprove of it'. Someone might accept the factual component – that factory farming causes suffering – but reject the use of the concept 'cruel' because they reject the attitude expressed about the practice.

The choice between realist and non-realist positions carries implications for a number of other meta-ethical questions about the status of ethical utterances. Are ethical judgements open to rational justification? Could we expect all rational agents to converge in their moral judgements? How are ethical judgements connected with actions? If one accepts a moral judgement, is one necessarily motivated to act upon it? What is the relationship between general ethical concepts – good, bad, right and wrong – and particular concepts – courageous, cowardly, kind, cruel, just, unjust and the like?

Discussions in environmental ethics are enmeshed in meta-ethical controversies. If environmental ethics is primarily concerned with substantive issues, why should this be so? What relevance do such meta-ethical disputes have for environmental ethics? The answer is that, where the relationship of humans to a non-human world is concerned, the pull of some form of realism about values has seemed to be particularly strong. Is the pull towards realism in environmental ethics a temptation to be resisted or one to which we should yield? To answer one way or another is to take a position on a general issue about the relation between meta-ethics and substantive issues in normative ethics. There is a longstanding view that the two spheres of philosophical discussion should be kept separate (Mackie 1977: 16). On this view, given that environmental ethics concerns substantive ethical issues, the excursion into meta-ethics is indicative of logical confusion. Alternatively, such excursions might give additional reasons for being sceptical about the view that meta-ethical and normative ethics can be kept separate (von Wright 1963: ch.1).

Intrinsic value

One of the main sources of the realist pull in environmental ethics has been the claim that to hold an environmental ethic is to hold that beings and states of affairs in the non-human world have intrinsic value. This claim is taken to distinguish 'deep' or 'biocentric' ethical theory from their more traditional 'shallow' and 'anthropocentric' counterparts. The term 'intrinsic value' however has a variety of senses, and many arguments on environmental ethics suffer from conflating them. It is worth starting consideration of the claim by making some distinctions between different senses of the term.

In its first and most basic sense 'intrinsic value' is used in contrast with the concept of instrumental value. 'Intrinsic value' is used as a synonym for 'non-instrumental value'. Objects, activities and states of affair have instrumental value insofar as they are a means to some other end. They have intrinsic value if they are ends in themselves. It is a well-rehearsed point that, under pain of an infinite regress, not everything can have only instrumental value. There must be some objects, activities and states that have intrinsic value. However, this concept of non-instrumental value is itself complex. It is sometimes predicated of objects, and sometimes also of states, or activities that an agent pursues or aims at. Of activities, one might say of a person who climbs mountains or studies the behaviour of birds that he indulges in these activities for their own sake. But the person might also be said to value the objects of these activities, the mountains or birds, for their own sake. Or he might be said to admire states of these objects for their own sake – the beauty of mountains or the complexity of a bird's behaviour. These activities, objects and states are said to be ends in themselves for the person.

The use of 'non-instrumental value' in this first sense is distinct from that employed, for example, by Kant when he claims that persons are ends in themselves, as we mentioned in previous chapters. To assert that a being is an end in itself in this Kantian sense is to assert that it has 'moral standing', which is to say that it counts morally in its own right for purposes of ethical assessment. As we noted in the last chapter a central move in much environmental ethics has been to extend moral standing beyond persons. To say that elephants, wolves, and even plants matter, in the sense that their good must be considered in making ethical choices, is to assign ethical standing to them. On the other hand, to say that one values the climbing of mountains, the beauty of mountains, or the mountains themselves for their own sake need not involve the ascription of any such ethical standing. However, while the notions are distinct, there is a prima facie plausible claim to be made about the relation between them. It is at least plausible to claim that if y is of value to x, and x has ethical standing, then there is a prima facie ethical duty for an agent not to deprive x of y.

'Intrinsic value' is also used in a third sense, in a contrast with 'extrinsic value', to refer to the value an object has solely in virtue of its intrinsic properties, that is its non-relational properties. Dampness is an intrinsic property of wetlands, for example, whereas their being endangered is extrinsic. The concept is thus employed by G. E. Moore: 'To say a kind of value is 'intrinsic' means merely that the question whether a thing possesses it, and to what degree it possesses it, depends solely on the intrinsic nature of the thing in question' (Moore 1922: 260).

Finally, 'intrinsic value' is also used as a synonym for 'objective value', that is, value that an object possesses independently of its being valued by any agent (Mackie 1997: 15). If 'intrinsic value' is used in this sense, then to claim that non-human beings have intrinsic value is not to make an ethical but rather a meta-ethical claim. It is to make a claim about the status of the value that they have – to assert a realist position about values.

If an environmental ethic is taken to be a substantive ethical position, according to which some non-human beings have intrinsic value, then 'intrinsic value' is being used in one of the first two senses: it is to hold that non-human beings are not simply of value as a means to human ends, but are ends in themselves, either in the sense of being valued for their own sake, or more strongly, in the sense of their having ethical standing. Arne Naess, the founder of 'deep ecology', one of the early distinguishable versions of environmental ethics, puts the point this way: 'The well-being of non-human life on Earth has value in itself. This value is independent of any instrumental usefulness for limited human purposes' (Naess 1984: 266). However, it might be claimed that to hold a defensible substantive ethical position about the environment, one needs to be committed to a particular meta-ethical position – the view that they also have intrinsic value in the third and, especially, fourth senses. This is a contention that we shall now proceed to discuss.

Is the rejection of meta-ethical realism compatible with an environmental ethic?

In much of the literature on environmental ethics the different senses of 'intrinsic value' are used interchangeably. In particular, intrinsic value understood substantively as non-instrumental value or as ethical standing is often conflated with intrinsic value understood meta-ethically as objective value – value independent of any valuer. Hence there is a widespread assumption that the rejection of a realist meta-ethics entails that non-humans can have only instrumental value. The assumption operates both ways. On the one hand those who claim that items in the non-human world have intrinsic value believe themselves to be committed to a realist view of values. On the other hand those who regard a realist view of values as indefensible infer that the non-human world can contain nothing of intrinsic value.

However, the claim that versions of a substantive environmental ethic are incompatible with the rejection of a realist meta-ethic is mistaken. In particular, the rejection of a realist meta-ethic does not commit one to the view that non-humans have only instrumental value. The apparent plausibility of the assumption that it does is founded on a confusion between the sources of value and the objects of value. An expressivist can be said to claim that the only sources of value are the evaluative attitudes of humans. But this does not entail that the only ultimate objects of value are the states of human beings. Likewise, to hold a realist view of the source of value according to which the value of an entity does not depend on the attitudes of valuers, is compatible with a thoroughly anthropocentric view of the object of value – that the only things which do in fact have value are humans and their states. (Other issues involving anthropocentrism are explored in Norton 1984; Hargrove 1992; and Katz 1999.)

To expand, consider the expressivist meta-ethic. Evaluative utterances express a speaker's attitudes. They state no facts. Within the expressivist tradition

Stevenson provides a clear account of intrinsic value. Intrinsic value is defined as non-instrumental value: 'intrinsically good' is roughly synonymous with 'good for its own sake, as an end, as distinct from good as a means to something else' (Stevenson 1944). Stevenson then offers the following account of what it is to say something has intrinsic value: 'X is intrinsically good' asserts that the speaker approves of X intrinsically, and acts emotively to make the hearer or hearers likewise approve of X intrinsically' (Stevenson 1944: 178). There are no reasons why the expressivist should not fill the X place by entities and states of the non-human world. There is nothing in the expressivist's meta-ethical position that precludes her holding basic attitudes that are not anthropocentric but instead much broader, or 'biocentric', and focused on the value of all living things. She can therefore readily hold that non-humans have ethical standing. There is no reason why the expressivist must assume either an egoist or humanist position. There is no reason why she must assign non-instrumental value only to herself, other humans, and their respective states and activities.

It might be objected, however, that there are other difficulties in holding an expressivist meta-ethic together with an environmental ethic which extends respect or rights to non-humans. In making humans the source of all value, the expressivist is committed to the view that a world without humans contains nothing of value. Hence, while nothing logically precludes the expressivist assigning non-instrumental value to objects in a world which contains no humans, it undermines some of the considerations that have led to the belief in the need to assign such value. For example, it is not compatible with the last man argument. The argument runs thus: if non-humans only have instrumental value, then the hypothetical last man whose last act is to destroy a forest would have done no wrong; the last man does do wrong; hence it is false that non-humans only have instrumental value. However, given an expressivist account of value the last man does no wrong, since a world without humans is without value.

This objection, however, still confuses the source and object of value. There is nothing in expressivism that forces the expressivist to confine the objects of her attitudes to those that exist at the time at which she expresses them. Her moral utterances might express attitudes towards events and states of affairs over periods in which she no longer exists – she might express her preference that her great-grandchildren live in a world without poverty; over periods in which humans no longer exist – she might express her preference that rain forests exist after the extinction of the human species and hence deplore the vandalism of the last man; and over different possible worlds – she might concur with Leibniz that this world is the best of all possible worlds, or, in her despair at the destructiveness of humans, express the attitude that it would have been better had humans never existed and hence a preference for a possible world in which humans never came into existence. That humans are the source of value is compatible with their expressing normative attitudes about worlds which they do not inhabit.

While the rejection of realism does not rule out non-humans having intrinsic value, neither do realist positions rule it in. To claim that moral utterances have a truth value is not to specify which utterances are true. The realist can hold that the moral facts are such that only the states of humans possess value in themselves; everything else has only instrumental value. Indeed a common view of ethical realists earlier this century was that only states of conscious beings have intrinsic value – a view based on the ground that any world without a mind would contain nothing good in itself.

What are the relations of intrinsic value in the sense of non-instrumental value and objective value, respectively, to the third, Moorean, sense of intrinsic value – to the value an object has that 'depends solely on the intrinsic nature of the thing in question' (Moore 1922: 260)? The intrinsic properties of an object are its non-relational properties. Many of the properties that are central to environmental valuation – rarity, species richness, biodiversity – are non-intrinsic in this Moorean sense. For example, rarity is an irreducibly relational property that cannot be characterised without reference to other objects, and in practical concern about the environment a special status is often ascribed to entities that are rare. The preservation of endangered species of flora and fauna and of threatened habitats and ecological systems is a major practical environmental problem. It has been argued that such value can have no place in an environmental ethic which holds non-humans to have intrinsic value. However, such arguments rely upon an equivocation between 'intrinsic value' in its Moorean sense and 'intrinsic value' used as a synonym for non-instrumental value. Thus, while it may be true that if an object has only instrumental value it cannot have intrinsic value in the Moorean sense, for instrumental value is necessarily value predicated on a relational property of an object, it is false that an object of non-instrumental value is necessarily also of intrinsic value in the Moorean sense. We might value an object in virtue of its relational properties, for example its rarity, without thereby seeing it as having only instrumental value for human satisfactions.

If there is value that 'depends solely on the intrinsic nature of the thing in question' then is it the case that we must embrace realism? If an object has value only in virtue of its intrinsic nature, does it follow that it has value independently of human valuations? The answer depends on the interpretation given to the phrases 'depends solely on' and 'only in virtue of'. If these are interpreted to exclude human evaluation and desires, as Moore intended, then the answer to both questions is immediately 'yes' – to have intrinsic value would be to have objective value. However, there is a natural non-realist reading of the same phrases. The non-realist can talk of the valuing agent assigning value to objects solely in virtue of their intrinsic natures. Hence, given a liberal interpretation of the phrases, a non-realist can still hold that some objects have intrinsic value in the Moorean sense.

The upshot of the discussion of this section is to re-affirm a traditional view – that meta-ethical commitments are logically independent of ethical ones. However, in the realm of environmental ethics it is a view that needs to be re-affirmed. No specific meta-ethical position is required by any specific environmental ethic. In particular, one can hold any such ethic and deny realism. However, this is not to say that there might not be other reasons for holding a realist account of ethics and that some of these reasons might appear particularly pertinent when considering evaluative statements about non-humans. The realist pull might still be a rational one.

Objective value and the flourishing of living things

We have argued that the claim that nature has intrinsic value in the sense of non-instrumental value does not commit one to a realist meta-ethics. However, in doing so we left open the question as to whether there might be other reasons particularly pertinent in the field of environmental ethics that would lead us to hold a realist account of value. Is there anything about evaluations of the environment that make the case for realism especially compelling?

Part of the pull for strong realism in the environmental sphere lies in a broadly Aristotelian observation that we have already discussed in the previous chapter, that there is a sense in which we can talk of what it is for natural entities to flourish, and what is good and bad for them, without this being dependent upon human interests (Attfield 1987; Taylor 1986; Rolston 1988: ch. 3). Thus consider again the farmer's use of the phrase 'x is good for weeds'. As we noted in the last chapter the term 'good for' can be understood in two distinct ways. First it can be used to refer to what is instrumentally good for the farmer – what is conducive to the destruction of weeds, as in 'Roundup is good for weeds'. Second it can be used to refer to what is instrumentally good for weeds – what is conducive to their flourishing, as in 'warm wet winters are good for weeds'. The possibility of using 'good for' in this second sense is due to the fact that living things are the sorts of things that can flourish or be injured (von Wright 1963: ch. 3). In consequence they have their own goods. These goods are independent of human interests. More significantly here, they are independent of any tendency they might have to produce in human observers feelings of approval or disapproval. A living thing can be said to flourish if it develops those characteristics which are normal to the species to which it belongs in the normal conditions for that species. Correspondingly, the truth of statements about what is good for a living thing, what is conducive to its flourishing, depends on no essential reference to human observers. The use of these evaluative terms in a biological context does then provide a good reason for holding that some evaluative properties are real properties. Their use does tell for some kind of realism about the use of the terms 'good' and 'goods' when these are applied to the non-human natural world.

If it is the case, then, that individual living things and/or the collective entities of which they are members can be said to have their own goods, then there are grounds for some kind of realism about some uses of the term 'good'. However, for reasons we outlined in the last chapter this leaves open the question whether the existence of such goods entails any human obligations, and hence whether it provides any argument for realism about specifically ethical goods. It is possible to talk in an objective sense of what constitute the goods of entities, without making any claims that these ought to be realised. For example, the farmer may know what it is for weeds to flourish, recognise that they have their own goods, and have a practical knowledge of what is good for them. No moral injunction follows. 'Y is a good of X' does not immediately entail 'Y ought to be realised' (Taylor 1986: 71–72). This gap clearly raises problems for environmental ethics. The existence of objective goods was promising precisely because it appeared to show that items in the non-human world were objects of proper moral concern. In the last chapter we examined various arguments that attempted to bridge the gap by extending traditional consequentialist and deontological theories outwards to include consideration of those goods, and suggested that they face serious problems.

How then can we show that the goods of non-humans ought to count in our moral considerations? We doubt the possibility of finding an argument that will compel assent and that is entirely unrelated to particular human responses and relations to the non-human world and the role these play in flourishing human lives. Accordingly a more promising option, in our view, is to look in the direction that we have been hinting at throughout the first part of this book. Human beings, like other entities, have goods constitutive of their flourishing, and correspondingly other goods instrumental to that flourishing. The flourishing of many other living things ought to be promoted because care for that flourishing, and the meaningful relationships with those other living things of which this care stands as an expression, is constitutive of our own flourishing. While, superficially, the approach might seem narrowly anthropocentric, it is not so in any objectionable sense. Friendship, which is the paradigm of a meaningful relationship, requires that we care for others for their own sake, even at some cost to ourselves and certainly not merely for the pleasures or profits they might bring. This is compatible with friendship being constitutive of a flourishing life. Given the kind of beings we are, a person without friends is likely to be leading an unhappy existence. On similar lines it can be argued that for at very least a large number of individual living things and biological collectives, we should recognise and promote their flourishing as an end itself. Such care for the natural world is constitutive of a flourishing human life. The best human life is one that includes an awareness of and practical concern with the goods of entities in the non-human world.

Environmental ethics through thick and thin

A feature of a great deal of theorising in environmental ethics, of which the search for 'intrinsic value' is typical, is that it loses sight of what moves environmental concern. There is a stark contrast between the richness of the normative vocabulary that informs our appraisal of the environments with which we live and the austerity of the vocabulary that environmental philosophers employ to theorise about it. As we noted in chapter 5, our appraisals of non-human nature call upon a range of normative vocabularies. For example, we can talk of cruelty inflicted on fellow creatures, of the vandalism involved in the wanton destruction of places rich in wildlife and beauty, of the pride and hubris exhibited in the belief that the world can be mastered and humanised, of our lack of a sense of humility in the midst of a natural world that came before and will continue beyond us. We possess a rich aesthetic vocabulary to talk of the tones, forms, sounds and textures of the natural world, the evocative quality of landscapes, the moods of nature. From the biological and ecological sciences come concepts that have normative significance, such as biodiversity, species richness, integrity, and fragility.

What, it might be asked, has all this rich vocabulary to do with the claim at the heart of much contemporary environmental ethics, that 'nature has intrinsic value'? References to intrinsic value only have power insofar as they call upon more specific reason-giving concepts, and corresponding claims about the ways in which natural objects are a source of wonder, the sense of proportion they invoke in us of our place within a wider history, the care we feel called upon to give as we develop our understanding of the lives of our fellow creatures, the diversity of forms of life to which we respond, and so on. Robbed of that more specific content one is left with concepts adrift that lend themselves to the kind of abstract metaphysics of value often to be found in environmental philosophy. The rich language that we employ in our discussions of environmental matters calls upon what are called 'thick' normative concepts. Thick normative concepts are specific reason-giving concepts, concepts like cruel, kind, just and unjust. They contrast with 'thin' concepts, general normative concepts like right, wrong, good or bad, or the favourites of environmental philosophy – 'has value' or 'lacks value' (Williams 1985).

A feature of thick concepts is that they are both descriptive and evaluative: to say that a farming practice is cruel is both to characterise and appraise the practice. As we noted in the opening section of this chapter, the non-realist responds to this apparent feature of thick concepts by prising apart the evaluative and descriptive components. We can analyse them as a conjunction of a factual component that does the describing and an evaluative component which is captured by the thin concept. So to say 'the practice is cruel' is to say something like: 'the practice involves the intentional infliction of suffering and the practice is bad'. The evaluative component is then given a non-realist reading – it serves merely to express attitudes or preferences. Now the realist might respond to this argument by accepting that thick concepts can be reduced to thin concepts in this way, but give a realist rather than expressivist account of the thin concept. However, another response might be to deny that thick concepts can be reduced to thin. The attempt to prise apart descriptive and evaluative components of thick concepts is not possible, since the descriptive content of the concept, its extension, is in part determined by the evaluative content. Only someone who understood the evaluative point of calling an act cruel or a mountain beautiful would know how to continue to use the concepts in new cases.

This response opens up the ground for a particular form of realism about ethical concepts that, unlike those we have considered thus far, does not divorce ethical responses from our human sensibilities and attitudes. Thus it might be argued that properties like 'cruelty' are real properties, specifically that: (i) our judgements track the properties; (ii) we can make mistakes; and (iii) claims that acts are cruel, or not cruel, are true or false. However, at the same time, they cannot be adequately characterised without reference to particular kinds of human evaluative responses to the world. Hence the feature of thick concepts noted earlier, that they are both action guiding and descriptive of states of the world. To characterise particular human acts as acts of cruelty is both to make a claim about the nature of those acts which is true or false, and to appraise them and offer reasons to oppose them. There do appear then to be grounds for a form of realism that stays at the level of thick ethical concepts.

Is a realism of thick concepts defensible? There have been a number of objections to giving priority to thick ethical concepts. One worry is that specific reason-giving concepts are culturally local – the intelligibility of the concepts relies on the particular practices of particular cultures. Hence, the objection goes, there is no possibility of a universal ethical language that is thick. That this is the case also militates against realism about ethical utterances, for one mark of truth for the realist is that it is that upon which reasoned judgements converge: if thick concepts are local, we cannot assume the possibility that reasoned judgements will converge in this way (Williams 1985: ch. 8). This meta-ethical point has additional significance for environmental problems since it can be argued that these are global and hence require an ethical language that crosses cultures. Any such language, the argument goes, is necessarily thin. Hence, the appeal to thin cosmopolitan concepts that is a feature of environmental philosophy is to be welcomed. Moreover, it might be argued that such cosmopolitanism in philosophy is part of the nature of its enterprise. It relies on the possibility of standing outside our own ethical practices and formulating sceptical questions about them – and for this, it might be added, we need a thin language to formulate criticism. We must be able to ask the question 'but is x good?' – for example, 'but is humility before nature good?'.

These arguments, however, assume that specific reason-giving concepts must be tied to culturally local contexts. But there is no reason to assume that this is true. In particular, there is no reason to assume in advance that thick ethical concepts cannot be universally shared while remaining open to local specifications. These might include concepts that characterise our relations to nature. This possibility is characteristic of an Aristotelian account of ethics, which holds that there are features we share as human beings that define what it is to lead a flourishing human life, and that characterisations of a good life will employ thick concepts, most notably the virtue concepts. Neither does the use of thick concepts rule out theoretical reflection and general principles in ethics. There is no reason to assume that critical theoretical reflection on our practices must be, or even can be, adequately undertaken using the thin ethical concepts through which much recent environmental ethics has been articulated. To ask 'Is humility before nature good?' is not to ask whether it has some property of goodness, but to raise questions about the relationship of such humility to other evaluative claims we might make – for example, about its compatibility with other admirable human accomplishments. True, the form of realism outlined here requires further elaboration and defence. However, it is more promising than the more popular realist position in environmental ethics which looks for values that exist independent of all human responses to the world.

The position we will defend in the rest of this book is one that attempts to call upon the rich normative vocabulary that starts from what David Wiggins calls 'the human scale of values' (Wiggins 2000). We reject the idea that, as we saw at the start of this chapter, has motivated much environmental ethics, and which calls for a 'new ethic' that needs itself to be grounded in metaphysical claims about value. We will be concerned with the place in our human lives of our relations to the environments which we inhabit. As we noted at the outset of this book, environments and the objects and beings they contain matter to us, and have meaning for us, in different ways. We live from them – they are the means to our existence. We live in them – they are our homes and familiar places in which everyday life takes place and draws its meaning, and in which personal and social histories are embodied. We live with them – our lives take place against the backdrop of a natural world that existed before us and will continue to exist beyond the life of the last human, a world that we enter and to which awe and wonder are appropriate responses. It is from our place within these various relations to the world that reflection needs to begin. In part three of this book we will begin to look at the way in which this approach might be developed in response to problems of biodiversity and sustainability. However, before doing so we will consider in the next chapter what sense we can make of talk of the value of nature.

8 Nature and the natural

'Nature,' 'natural,' and the group of words derived from them, or allied to them in etymology, have at all times filled a great place in the thoughts and taken a strong hold on the feelings of mankind . . . The words have . . . become entangled in so many foreign associations, mostly of a very powerful and tenacious character, that they have come to excite, and to be the symbols of, feelings which their original meaning will by no means justify, and which have made them one of the most copious sources of false taste, false philosophy, false morality, and even bad law . . .

(J. S. Mill, *On Nature*)

Valuing the 'natural'

A central claim of those who advocate the need for a new environmental ethic has been that nature has intrinsic value. In the last chapter we considered the meta-ethical debates that this claim has engendered. In this chapter we will consider the claim itself in more detail. In particular we will examine the status of one strong version of the claim, that at least part of the basis for our concern with the natural world is that we value what is natural as such (i.e. we value it just because it is natural). 'Naturalness' is itself taken to be a source of value. As Goodin puts it: 'According to the distinctively [green theory of value] . . . what it is that makes natural resources valuable is their very naturalness' (Goodin 1992: 26–27). Correspondingly what many environmentalists take to be at stake in our environmental crisis is the disappearance of this natural world. A strong and eloquent statement of this idea can be found in Bill McKibben's book *The End of Nature*:

An idea, a relationship, can go extinct just like an animal or a plant. The idea in this case is 'nature', the separate and wild province, the world apart from man to which he has adapted, under whose rules he was born and died. In the past we have spoiled and polluted parts of that nature, inflicted environmental 'damage' . . . We never thought that we had wrecked nature. Deep down, we never really thought we could: it was too big and too old.

Its forces – the wind, the rain, the sun – were too strong, too elemental. But, quite by accident, it turned out that the carbon dioxide and other gases we were producing in our pursuit of a better life – in pursuit of warm houses and eternal economic growth and agriculture so productive it would free most of us for other work – *could* alter the power of the sun, could increase its heat. And that increase *could* change the patterns of moisture and dryness, breed storms in new places, breed deserts. Those things may or may not have yet begun to happen, but it is too late to prevent them from happening. We have produced the carbon dioxide – we have ended nature. We have not ended rainfall or sunlight . . . .But the *meaning* of the wind, the sun, the rain – of nature – has already changed.

(McKibben 1990: 43–44)

McKibben mourns what he takes to be the passing of nature, but his lament raises many questions. Who or what is the 'nature' whose loss McKibben mourns, and what sense can we make of the 'loss' of nature as such? How adequate is McKibben's own conception of nature, in particular his identification of it with wilderness, 'the separate and wild province, the world apart from man'? And what sense can we make of the claim that something is valuable in virtue of being natural?

These are the questions we shall be examining in this chapter. We will examine the different senses in which one can talk of 'nature' and the 'natural', and examine attempts to argue that naturalness is a source of value. In doing so we will also consider the paradoxes that are sometimes taken to follow from what are described as attempts to 'restore' nature. We will argue that what does emerge from consideration of 'naturalness' as a value is the particular role that history and narrative play in our evaluative responses to the environments, beings and things around us. However, this role is not confined to things that are natural but applies as well to cultural landscapes and objects, and also to our relations to fellow human beings. We value a variety of entities as spatio-temporal particulars, as beings with a particular temporal history. We will explore this claim in more detail in the final part of the book.

The complexity of 'nature'

Some distinctions

David Hume writes of the word 'nature' that 'there is none more ambiguous and equivocal' (Hume 1978 [1739]: III.i.ii). The claim is one that has been repeated in recent discussion. Raymond Williams, for example, asserts that 'Nature is perhaps the most complex word in the language' (Williams 1976: 184). That complexity is a source of much argument in environmental philosophy. What is the 'nature' that environmentalists aim to defend and protect and whose loss is mourned? Consider Richard North's dismissal of McKibben's worries in the following terms.

> I think it is always right for man to consider the sadness of his fallen condition. But we have been thrown out of paradise, that's all. All the other stuff – the *End of Nature* song – is quite wrong. For a start . . . man can never damage nature, because nature is a set of scientific facts by which man can live, or die. Man never had dominion over natural laws. What he had, and has, is a certain power to change the face of the earth, and now, fractionally, its climate.

> (North 1990)

How effective is such a response? In particular, are McKibben and North using the term 'nature' in the same sense and so are they really engaging with each other, or are they simply talking about different things?

To get some initial orientation it is worth considering Hume's attempts to disambiguate different senses of 'nature'. 'Nature', he notes, is used in a series of contrasts. It is sometimes opposed to the miraculous, sometimes to what is artificial, sometimes to the civil and sometimes to the rare and unusual. The last sense of the term that Hume refers to, in which the natural is used as a synonym for the common or usual, will not concern us here, although there may be more to the concept than is often assumed. The sense in which it is used as a contrast to the civil we will discuss further below. Of the contrast between nature and the miraculous Hume notes the following: 'If *nature* be opposed to miracles, not only the distinction betwixt vice and virtue is natural, but also every event, which has ever happened in the world, *excepting those miracles, on which our religion is founded*' (Hume 1978 [1739] III.i.ii: 474). Given what we know of his scepticism about reports about miracles, there are good reasons to suppose that the last phrase in this sentence should be understood as an ironical aside. Once one rejects the idea of the miraculous then 'every event, which has ever happened in the world' is natural. The point can be made a little more generally. The opposition of the natural and the miraculous can be understood as an aspect of the contrast between the natural and the supernatural. If one rejects the idea of the supernatural then everything is natural. However, the term 'nature' can also be understood in a narrower sense: '*nature* may also be opposed to artifice' (Hume 1978 [1739] III.i.ii: 474). It is this latter distinction that seems to be relevant for discussion of the value of nature. How is this distinction to be understood?

An influential answer to that question is offered by J. S. Mill:

> It thus appears that we must recognise at least two principal meanings in the word 'nature'. In one sense, it means all the powers existing in either

the outer or the inner world and everything which takes place by means of those powers. In another sense, it means, not everything which happens, but only what takes place without the agency, or without the voluntary and intentional agency, of man. This distinction is far from exhausting the ambiguities of the word; but it is the key to most of those on which important consequences depend . . .

(Mill 1874: 8–9)

Mill in this passage distinguishes two meanings of the word 'nature' that parallel the distinction drawn by Hume: first, a broader sense in which it refers to 'all the powers existing in either the outer or the inner world and everything which takes place by means of those powers' or as he puts it more pithily earlier in the essay, in which 'nature . . . is a collective name for all facts actual and possible' (Mill 1874: 6); second, a narrower sense in which it refers to what takes place 'without the voluntary and intentional agency of man'. The first sense of the term signifies roughly what is 'natural' as opposed to 'supernatural'. The second sense registers the contrast between the 'natural' and the 'artificial'.

Natural and artificial

If we consider the concept of the artificial we see that it is used in a number of different, though connected, senses:

1. In one sense it means 'phoney' or 'bogus'. We speak, for example, of 'artificial laughter'.
2. In another sense it means 'substitute'. We speak of an 'artificial limb', or 'artificial light'.
3. In yet another sense it means 'human-made'. We speak of an 'artificial lake'. The lake might be a substitute for a natural lake, but again, it might not.

This last sense is the one that concerns us, but it still needs some refinement. This can be seen by reflecting on the fact that human tears, for example, are a human product, even though they well up quite 'naturally' (unless they are so-called 'crocodile tears'). This suggests that in order to capture the precise meaning of the term 'artificial' we need to introduce the idea of contrivance, and this is precisely what Mill does with his reference to 'voluntary and intentional agency'. Something is artificial only if it is the result of a deliberate or intentional act. A further refinement is to distinguish between the *result* and the *aim* of a deliberate or intentional act. This distinction will be seen to have considerable importance when we reflect on the phenomenon of global warming, which is claimed by many to be the result, though it was not of course the original aim, of the accumulation of deliberate choices.

However, while something is artificial only if it is the result of a deliberate and intentional act, the claim does not quite work the other way around. It is not true that whatever is the result or aim of a deliberate or intentional act is artificial. Human beings themselves, for example, are often the result of deliberate or intentional acts of human procreation, but it would be odd to classify human beings as artefacts. We need to make a further distinction, therefore, between intentions which simply bring a thing into existence and intentions which determine and shape the nature of the thing – the properties that make it what it is. Consider, for example, the case of a highly cultivated rose or a genetically modified plant. If we graft a cutting, or sow the seed, of such a plant, what grows might be described as an 'artefact'. But if it were so described, what would make it an artefact would not be the fact that we grafted it or planted it, and in that sense caused it to come into existence, but rather the fact that human ingenuity would have gone into shaping the kind of plant that it is – The Queen of Denmark (an Alba rose), for example, or such and such a variety of maize. We can arrive, then, at the following working definition of the term 'artificial': Something is artificial if and only if it is what it is at least partly as the result of a deliberate or intentional act, usually involving the application of some art or skill. Something's being artificial is a matter of its nature being determined by a deliberate and intentional act. (For an account along similar lines, see Siipi 2003.) One of the best known applications of this distinction between 'natural' and 'artificial' is Charles Darwin's introduction of the terms to distinguish between the process that is largely responsible for evolution, which he calls natural selection, and the process that drives domestication and cultivation, which he calls artificial selection or 'selection by man'. We also find in Darwin a recognition of the further distinction between the aim and the result of deliberate action. He reserves the term 'unconscious selection' for the modification of varieties that is the *result* but not the *aim* of deliberate choice, and 'methodical selection' when the aim is indeed 'the modification of the breed'.

Clearly, however, there are differences of degree, and perhaps in kind, of artificiality, thus understood. Consider for example the development of new techniques for genetically modifying crops. One typical source of popular concern about such techniques is that they are 'against nature' or 'unnatural' (Eurobarometer 2000; Macnaghten 2004). And one typical response to that concern is that traditional techniques for the development of new varieties of crops by processes of artificial selection are themselves 'unnatural' and hence there is no difference in kind between new techniques of genetic modification and older agricultural techniques (Nuffield Council 1999). It could be argued, however, that this response hardly settles the question. The first generation of Mendelian artificial selection typically mimicked natural selection. It is precisely for this reason that Darwin could invoke processes of artificial selection to help explain natural selection. They generally involved forms of genetic modification that do not occur in nature for contingent reasons, such as spatial or temporal divergence, rather than because they involve processes of selection that could not in principle occur in the natural world. In contrast, contemporary techniques of genetic modification, in particular when they involve the crossing of species boundaries, sometimes involve processes of selection that we would not expect could occur in the absence of human intervention. An argument could be made then that there is a difference between selecting future lineages from among those that are possible or probable, as in the original experiments in breeding and crossing, and determining which lineages shall be possible, as in some of the new techniques of genetic modification and in much modern, hi-tech versions of conventional breeding. The difference reflects, and is possibly as significant as, the one we drew earlier between being responsible for bringing something into existence and being responsible for shaping the kind of thing that it is. Whether this distinction can settle the question of whether genetic modification is 'unnatural', and whether a successful charge that it is unnatural should count in an assessment of that technology is of course another set of issues entirely.

Let us return now to the distinction between two different senses of nature drawn by Hume and Mill, and ask what we are to make of North's response to McKibben. It is not clear that North's criticism is effective. When North claims that humans 'can never damage nature', he is clearly referring to nature in the first of the two senses distinguished by Hume and Mill in which it refers to what Mill calls 'all facts actual and possible'. Whatever humans do, they are always subject to the laws of nature and can never control them. But when McKibben talks of the 'end of nature', he seems clearly to mean nature in the second sense distinguished by Hume and Mill in which it is contrasted with the artificial. He is concerned about how much of what was natural is becoming artefactual, or at least is being altered by artificial and especially technological processes. First, then, North is failing to address the claim that McKibben is making. Second, he is failing to address McKibben's argument that the changes which humans are now causing are of a different order from those they have caused in the past.

That said, this argument assumes that there is a significant distinction to be drawn between what humans do intentionally and everything else that happens. There are grounds however for dissatisfaction with such a distinction. First, if other animals are capable of intentional acts, it may seem arbitrary to single out the intentional acts of humans as of special significance. It is true that we normally think of humans as the only creatures capable of producing artefacts. In principle, however, any creature that is capable of learning should probably also be judged capable of producing an 'artefact'. Thus, a beaver's dam should almost certainly be counted as an artefact, whereas a spider's web probably should not. Second, and more importantly, to picture nature as the world from which intentional human acts have been abstracted may seem unreal, given that intentional human agents are as much products of nature as are sunflowers and seahorses. This is a tension that indeed goes to the heart of our environmental predicament, and is a key challenge that any account of environmental value must recognise and meet. Third, the major sources of concern to which McKibben draws attention, global warming, for example, or the damage to the ozone layer produced by CFCs, are in fact side-effects of what humans have done intentionally. In terms of our earlier distinction, they were the unanticipated result, not the aim, of intentional actions. Consequently it becomes problematic where exactly they belong within the crude dichotomy of what humans do intentionally, and everything else that happens.

Natural and cultural

As Mill notes, however, the distinction between the 'natural' and 'artificial' does not exhaust the contrasts that define the different senses of 'natural' and 'nature'. The terms are also used to draw a variety of other contrasts such as those between the 'natural' and the 'social' or the 'cultural'. Thus, for example, distinctions are drawn between the 'social sciences' and the 'natural sciences'. Those who want to deny that there is a hard distinction to be made here sometimes talk of taking a 'naturalistic perspective' in the social sciences. Hume similarly notes that *'natural* is also sometimes opposed to *civil'* (Hume 1978 [1739] III. i.ii: 475). This contrast picks up a central distinction in early modern political theory. Thus, from Hobbes through to Rousseau a distinction was drawn between the 'state of nature' and 'civil society', in which political and social institutions of a particular kind were said to exist. For different theorists such terms do different work. In Hobbes, the contrast indicates the awful fate that would befall us in the absence of political institutions. The virtues of a civil society lie in their allowing us to avoid the state of nature that would befall us in its absence. For Rousseau the distinction is used to opposite effect. Civil society is taken to be a realm of artificiality disconnecting us from our original, benign state of nature. This latter theme is found in the work of the romantic poets as well: the natural is where we find what is authentic and right, in contrast to the social artificialities and contrivances of the social world. Of this latter theme Raymond Williams notes:

> One of the most powerful uses of nature since the eighteenth century has been this selective sense of goodness and innocence. Nature has meant the 'countryside', the 'unspoiled places', plants and creatures other than man. The use is especially current in contrasts between town and county: nature is what man has not made, though if he made it long enough ago – a hedgerow or a desert – it will usually be included as natural. Nature-lover and nature poetry date from this phase.

(Williams 1976: 188)

Nature as wilderness

This use of 'natural' as what is 'rural' is however one that is likely to be confined now only to Europe. In the 'new worlds' the term 'natural' has tended to be used much more starkly to refer to 'wilderness' marked by, in John Muir's words, the absence of 'all . . . marks of man's work'. (For a dissenting voice, see Cronon 1996.) Hence what is rural, insofar as this embraces 'domesticated' and 'cultivated' landscapes, comes to be sharply contrasted with what is natural, embracing as it does only those landscapes that we might call 'wilderness'. This is the sense of nature which McKibben is celebrating in the passage quoted above. It is in this spirit that he writes of nature as a 'separate and wild province, the world apart from man to which he has adapted'.

However, by no means all marks of human activity are there by contrivance. Hence, in the spirit of a point we made earlier (pp. 128ff) – that the artificial implies the idea of contrivance – we must be careful to preserve a distinction between this identification of nature with 'wilderness', understood as that which is free of all marks of human activity, and the notion of nature that contrasts simply with what is artificial. As David Wiggins puts it, 'nature . . . understood as that which is free of all traces of our interventions' needs to distinguished from nature as 'that which has not been entirely instrumentalized by human artifice . . . ' (Wiggins 2000: 10). There is little in Europe for example that lacks the marks of human activity. However, it would be wrong to conclude that the concept of nature in the sense that contrasts with what is artificial lacks application in this context.

What are we to make of the identification of 'nature' with 'wilderness'? An initial question to be asked here is whether nature as wilderness has the same significance – or indeed picks out the same thing – regardless of a person's social and cultural background. It certainly seems as if a person's ideas about nature are likely to be affected by their cultural circumstances. For a North American like McKibben, nature is likely to signify wilderness as it is found in federally designated wilderness areas in the US, like Glacier National Park; Australians conceptualise 'the bush' in similar terms. On the other hand, as we have just noted, a European is most likely to think of the rural environment – the country-side – as constituting 'nature'. More significantly, indigenous populations of America and Australia had different conceptions of those places designated now as wilderness. For those who lived in the 'wilderness' of the new world, this was not a wilderness but home, and a cultivated home at that. What immigrants from the old world encountered in the new was another people's old world, another people's home.

The failure to recognise the previous aboriginal transformations of landscapes has in fact been a source of problems in the treatment of the ecology of the new world. In particular, in both Australia and the United States, the understanding of nature as a primitive wilderness has led to a failure to appreciate the ecological impact of native land management practices, especially those involving burning. Consider the history of the management of one of the great symbols of American wilderness, Yosemite National Park. In the influential 1963 report of the Starker Leopold Committee, *Wildlife Management in the National Parks*, commissioned for the US Department of the Interior to help set national policy, we find the following statement of objectives for parks: 'As a primary goal we would recommend that the biotic associations within each park be maintained, or where necessary be recreated, as nearly as possible in the condition that prevailed when the area was first visited by the white man. A national park should represent a vignette of primitive America' (cited in Runte 1987: 198–199). What was that condition? The first white visitors represent the area thus:

> When the forty niners poured over the Sierra Nevada into California, those who kept diaries spoke almost to a man of the wide-space columns of mature trees that grew on the lower western slope in gigantic magnificence. The ground was a grass parkland, in springtime carpeted with wildflowers. Deer and bears were abundant.
>
> (ibid.: 205)

However, this 'original' and 'primitive' state was not a wilderness but a cultural landscape with its own history. The 'grass parkland' was the result of agri-cultural practices of the native Americans who had used fire to promote pasture for game, black oak for acorns, and so on for centuries. After the Ahwahneeche Indians were driven from their lands in Yosemite by a military expedition in 1851, 'Indian style' burning techniques were discontinued and fire-suppression controls were introduced. The consequence was the decline in meadowlands under increasing areas of bush. When Totuya, the granddaughter of chief Tanaya and survivor of the Ahwahneeche Indians who had been evicted from the valley, returned in 1929, she remarked of the landscape she found: 'Too dirty; too much bushy'. It was not just the landscape that had changed. In the Giant Sequoia groves the growth of litter on the forest floor – consisting of dead branches and competitive vegetation – inhibited the growth of new Sequoia, and threatened more destructive fires (Olwig 1996). Following the Leopold report, both cutting and burning were used to 'restore' Yosemite back to its 'primitive' state.

Thus we see that reference to wilderness suppresses one part of the story that can be told of the landscape. The non-European native occupants of the land are themselves treated as part of the 'natural scheme', of the 'wilderness'. Their history as dwellers in a landscape which embodies their own cultural history is made invisible. Moreover, such language also disguises the way in which the history of the landscape is being frozen at a particular point in time. Again, from the Leopold report, 'The goal of managing the national parks and monuments should be to preserve, or where necessary recreate, the ecologic scene as viewed by the first European visitors' (ibid.: 200). To refer to 'natural' or 'wilderness' states avoids the obvious question, Why choose that moment to freeze the landscape? There are equally obvious answers to that question, but they have more to do with the attempt to create an American national culture around such places than with ecological considerations. Politically, the use of the term 'wilderness' to describe areas of the globe can be much less innocent than some of its proponents assume. Historically, it has the effect of disguising the fact that colonised land was the home of someone else: it is part of the designation of the land as *terra nullis*, unowned land. Where indigenous populations are recognised, the designation of their home as 'wilderness' tends to promote an image of them as part of the fauna rather than as fellow human beings. Furthermore, it seems to us clear that the appropriate response to a recognition of the role of indigenous peoples in shaping the US and Australasian landscapes is surely *not* to infer that these landscapes do not have the value that we thought they had. Rather it should be to develop a different account of the value that they do have.

The value of natural things

We now return to the question of whether natural things can be said to have a value just because they are natural. Someone who believes very strongly that this is so is the American philosopher Eric Katz. To make such a case, however, it is not enough simply to affirm that natural things have value. It is necessary to say what it is about natural things that gives them their value, and this, indeed, is what Katz tries to do, using a broadly Kantian rationale. Katz's discussion turns on the distinction between the natural and artificial. Katz suggests that if it is possible to conceive of things in nature as moral subjects, and if we follow Kant's argument that things which are moral subjects should never be treated as mere means to further our own ends but as ends-in-themselves that should be respected, then natural things should be treated as ends-in-themselves, while artefacts can be treated as mere means to further human ends. Following from this analysis, Katz derives a moral claim that we should not intervene in nature in such a way that we come to treat it like an artefact:

> This consideration of Kantian moral concepts *suggests* that two crucial notions in the development of an ethical environmental policy are the Kantian ideal of 'autonomy', and its moral opposite, domination. In analysing the value of natural organisms, Rolston writes: 'the values that attach to organisms result from their nonderivative, genuine autonomy . . . as spontaneous natural systems.' This is not true merely for organisms.

Complex holistic natural systems and communities also exhibit autonomy, in that they are independent from external design, purpose, and control. Even non-living natural entities, which do not, in themselves, develop, grow, or achieve self-realization, are essential components of autonomous natural systems. When humans intervene in nature, when we create artefacts or attempt to manage environmental systems (such as forests), we destroy that natural autonomy by imposing a system of domination . . .

(Katz 1993: 230)

A contrasting voice is that of Mill, who states the case, rather powerfully, in his *Essay on Nature* for rejecting the claim that 'being natural' is any kind of commendation at all:

In sober truth, nearly all the things which men are hanged or imprisoned for doing to one another are nature's every-day performances. Killing, the most criminal act recognised by human laws, Nature does once to every being that lives; and, in a large proportion of cases, after protracted tortures such as only the greatest monsters whom we read of ever purposely inflicted on their living fellow creatures. If, by an arbitrary reservation, we refuse to account anything murder but what abridges a certain term supposed to be allotted to human life, nature also does this to all but a small percentage of lives, and does it in all the modes, violent or insidious, in which the worst human beings take the lives of one another. Nature impales men, breaks them as if on the wheel, casts them to be devoured by wild beasts, burns them to death, crushes them with stones like the first Christian martyr, starves them with hunger, freezes them with cold, poisons them by the quick or slow venom of her exhalations, and has hundreds of other hideous deaths in reserve, such as the ingenious cruelty of a Nabis or a Domitian never surpassed. All this Nature does with the most supercilious disregard both of mercy and of justice, emptying her shafts upon the best and noblest indifferently with the meanest and worst.

(Mill 1874: 28–29)

Later in the same essay he concludes:

[T]he doctrine that man ought to follow nature, or, in other words, ought to make the spontaneous course of things the model of his voluntary actions, is equally irrational and immoral.

(Mill 1874: 64)

Katz and Mill clearly offer very contrasting views on the value of natural things, which in turn have very different policy implications. Katz holds that the crucial feature of the entities and systems that comprise nature, and what distinguishes them from artefacts, is their independence from human aims and goals. He identifies autonomy and self-realisation as pre-eminent values, and claims that natural entities and systems have value by virtue of exhibiting these characteristics. Hence, human policy towards the natural world should be limited to such intervention, alteration and management as is necessary for human self-realisation, but should stop short of domination. Applying these principles to forestry policy, he goes on to argue in his 1993 essay that even sustainable forestry oversteps such limits and is unnecessarily destructive of the moral value of natural forests.

Mill, on the other hand, while allowing that nature is impressive and even awe-inspiring, insists that its powers are also both hurtful and indifferent, and that 'man must wrest by force and ingenuity what little he can for his own use'. He seems, at least in this essay, to suggest that humans are entitled to do all in their power to ward off natural calamities and to alleviate and improve the human condition. He finds 'draining a pestilential marsh' no more reprehensible than 'curing the toothache or putting up an umbrella'.

There are clearly problems with the arguments of both Katz and Mill which deserve further exploration. With respect to Katz's argument one might wonder whether it makes sense to ascribe autonomy and self-realisation, in the same sense in which these terms are used to describe people, to nature and to natural processes. Terms such as autonomy and self-realisation, for example, imply the existence of individuals – that is, items that can be individuated, identified and re-identified over time. And with the exception of individual animals, it is hard to locate such entities in nature, without committing oneself to something approaching Frederick Clements's 'organismic' view of natural communities. Hence, Katz's position appears to depend upon a specific and highly contentious ecological theory. A second problem is this. In human terms autonomy and self-realisation are important because it matters to the individuals concerned if their aspirations and projects are thwarted or stunted. But it is difficult to carry this across to the case of natural processes. Perhaps it matters *that* a stand of trees is clear-cut. But does it matter *to* a stand of trees if it is clear-cut? If the terms can be used at all, do they have the same meaning when applied to natural processes? A third problem stems from the fact that autonomy and self-realisation are competitive goods. Can a clear line be drawn between human self-realisation and human domination of nature? Perhaps humans can only achieve self-realisation through such domination. In civil society, a great deal of human self-realisation depends upon the (often tacit) forbearance, cooperation, inhibition and self-control of others. It is not clear, on the other hand, that natural self-realisers are also natural forbearers. Certainly, forms of cooperation may emerge, but only over ecological, rather than cultural, time-scales.

Mill on the other hand appears to conflate evaluations of intentional human actions with those of natural processes that are not intentional. From the fact that an action would be bad if perpetrated by a human being, it does not follow that it is bad when perpetrated by non-human nature. Moreover, the fact that some natural events are harmful does not entail that 'naturalness' is not of value; any more than the fact that compassionate or honest people sometimes do harm entails that compassion or honesty is of no value. There is also an interesting tension in Mill's own work. It is worth comparing here his arguments in this passage from *On Nature* with the earlier passages from his *Political Economy* which we quoted in chapter 2 concerning the value of solitude in the presence of natural beauty.

Recall there that Mill decries a world bereft of land that is not domesticated or cultivated. Mill holds that there is value to be found in the products of 'the spontaneous activity of nature'. However, his arguments are different in kind from those offered by Katz. For Katz what is wrong with the control of the spontaneous activities of nature is that it inhibits the 'self-realisation and autonomy of nature' understood independently of the self-realisation of humans. For Mill, on the other hand, the self-realisation of humans requires that they be able to escape from other humans. It is not good for human beings to see everywhere the hand of human work. The existence of processes and objects that are independent of human intentionality is taken to be good for our own development as individuals and as a society. Whether that is the case and for what reason is itself an important question and one to which we will return on pp. 146ff.

Other accounts have been offered of what gives natural things their value. But they do not often provide an account of what gives natural things a value *by virtue of their being natural*. An example is the recent attempt by Richard Norman to provide an aesthetic foundation for natural value (Norman 2004). (Among earlier attempts are those of Thomas Hill Jr. (1983); Elliot Sober (1986); and John Lawton (1991).) Briefly, Norman's suggestion is that we look for the value of nature in the range of 'expressive' qualities that it exhibits, quali-ties that are at once aesthetic but also charged with meaning, and furthermore 'contribute to the meaning the world has for us' (Norman 2004: 23). As instances he cites being bleak, awe-inspiring, relentless or menacing. We do not deny, of course, that nature has these qualities to a high degree. Nor do we deny that Norman's account succeeds in capturing a non-instrumental view of nature: the 'uses' of a bleak terrain and relentless rain are limited. But we do deny that this kind of account can capture the value of nature as such. This is, first, because human constructions can share these same expressive qualities that he attributes to nature. The outline of a disused power-station, for example, might well rival a rocky crag in its menacing bleakness. Or a human-made dam might rival any number of natural waterfalls in its awe-inspiring and relentless power. The second reason is, because nature has many qualities that are not at all 'expressive', or at any rate, are not obviously sources of value in the same way. Norman places particular emphasis upon so-called 'sublime' qualities such as the bold, the barren and the bleak. Others have placed the emphasis on nature's beauty. But the truth is that many features of nature are at least disturbing, if not downright ugly, to the ordinary eye. Think for a moment of the heaving mass of maggots that swarm over the rotting carcass of a dead sheep. But more significant still perhaps are the vast tracts of nature that are simply mundane and monotonous. If we are to find value in nature as such, then we must find value in these too. To find value in *spectacular* nature is not the same as finding value in nature (Holland 2004).

A different approach again is to find ethical significance in the characteristics, structures and processes of nature that are disclosed by the study of ecology. Because of its ecological inspiration this approach tends to focus on items such as populations, species, habitats, communities and ecosystems, rather than on individuals. And the ethical significance tends to be framed in terms that mimic those that we apply to human welfare. Thus it is urged that the 'health' of natural communities, or of the 'land' (Leopold 1949), should command our respect, as should the 'goals' of natural systems (Johnson 1991), or the 'intentionality' of nature more generally (Plumwood 1998; see also Plumwood 1993).

We see several disadvantages in this approach. One is the amount of heaving, stretching and hauling that has to go on, in order to represent nature as amenable to the categories of conventional ethics – health, interests, purposes and the like. We are inclined to protest that if we value the natural world, it ought to be for what it is, rather than because it is like something else. Another is that such an approach makes the value of the natural world too much hostage to the fortunes of particular ecological theories which have been rapidly proposed and rejected in the fairly short history of the science. It ignores the fact that scientific concepts generally are always likely to be superseded, which means that ethical judgements built upon particular ecological descriptions are likely to prove fragile (Holland 1995a). Furthermore, there is the question of how far the project of building values upon ecological descriptions is likely to do justice to the appeal of the wild and disorderly. Certainly, many ecologists acknowledge the haphazard qualities of the natural world. Famously, and in opposition to Frederick Clements, Herbert Gleason insisted that plant communities were mere 'fortuitous juxtapositions' (Gleason 1936, cf. Gleason 1927: 311). Finally, although several of these items – species, for example, or communities – can be viewed historically, as we urge they should be, this aspect has been generally overlooked in so-called 'ecological' ethics. We return to this point below.

Nature conservation

Nature conservation is one of the primary expressions of environmental concern. We have devoted some attention to natural value because the justification behind all forms of nature protection is presumably the belief that, in conserving nature, we are conserving something of value. However, nature conservation, and associated projects such as nature restoration, present us with a fresh set of problems.

For the purpose of our present discussion we might broadly distinguish three related projects:

1. protecting nature from human incursion,
2. restoring natural features if they are damaged,
3. restitution for natural features that have been destroyed.

Thus, in protecting nature we conceive ourselves as protecting something of value; in restoring nature we conceive ourselves as restoring something of value; while restitution involves the attempt to make amends for a loss that has been caused by creating something of value that in some way makes up for what is lost.

A paradox?

However, in light of the definition of 'nature' that we have been assuming (the second of Mill's two senses), there appears to be something extremely paradoxical about the idea that we can protect or restore nature. It is analogous to the paradox that Mill found in the idea that we might adopt the injunction to 'follow nature' as a principle of action: following nature would have to be a deliberate act, and precisely for that reason could not count as following *nature*. There appears to be a similar paradox in the idea of attempting to protect or restore nature. The paradox is this. Nature, in the second of the two senses identified by Mill, is what exists outside of any intentional human intervention. Protection or restoration, on the other hand, requires intentional human intervention in order to put it into effect. So, how can it be possible to protect or restore, by intentional human agency, something that is supposed to be independent of intentional human agency?

It might be argued that the paradox of nature protection is more apparent than real. Richard Sylvan, for one, is quite scathing in his dismissal of the paradox, labelling it as the 'bottom of a barrel of rotten arguments' (Sylvan 1998). Robert Elliot concurs:

> The idea is that by placing boundaries around national parks, by actively discouraging grazing, trail-biking and the like, by prohibiting sand-mining, we are turning the wilderness into an artefact, that in some negative or indirect way we are creating an environment . . . But . . . what is significant

about wilderness is its causal continuity with the past. This is something that is not destroyed by demarcating an area and declaring it a national park. There is a distinction between the 'naturalness' of the wilderness and the means used to maintain and protect it. What remains within the park boundaries is, as it were, the real thing . . . There is a significant difference between preventing damage and repairing damage once it is done. That is the difference that leaves room for an argument in favour of a preservation policy over and above a restoration policy.

(Elliot 1995: 87)

The central move in Elliot's account is that preservation sustains the 'causal continuity' of a place with its past, whereas restoration does not. An initial question to ask of that account is whether the concept of 'causal continuity' per se is strong enough to do the work that Elliot wants it to do. In its weakest sense any development has causal continuity with the past including those involved in restoration in the examples to which Elliot particularly objects: bulldozers that flatten the land and diggers replanting it are all part of causal processes linking the past state with the present. What Elliot is after in the term 'causal continuity' is a particular kind of causal history linking past and future, one that allows us to differentiate the causal processes involved in protection from those involved in restoration. Consider, for example, the removal of feral animals or the eradication of non-indigenous plants. Do these count as 'protection' or 'restoration'?

Moreover, as we come to thicken up our account of what counts as 'causal continuity' one might question the claim that preservation policies that place boundaries around a national park and exclude human activity in order to create a 'wilderness' do count as sustaining causal continuity with the past of a place. Rather, such a policy can involve the exclusion of local populations for whom what is now a park was previously their home in a way that clearly creates discontinuities with the past. Consider for example the exclusion of the San people from their ancestral lands in the Kalahari on the grounds that their pastoral activities are now inconsistent with the status of those lands as a nature reserve (The *Guardian*, 5 March 2004). Or consider the following quotation from a local peasant no longer able to continue some of his traditional activities within the National Park of Sierra Nevada and Alpujurra granted biosphere status by UNESCO, and Natural Park status by the government of Andalusia:

[Miguel] pointed out the stonework he had done on the floor and lower parts of the wall which were all made from flat stones found in the Sierra. I asked him if he had done this all by himself and he said 'Yes, and look, this is nature' ('Si, y mira, esto es la naturaleza'), and he pointed firmly at the stone carved wall, and he repeated this action by pointing first in the direction of the Sierra [national park] before pointing at the wall again.

Then, stressed his point by saying: 'This is not nature, it is artificial (the Sierra), this (the wall) is nature' ('Eso no es la naturaleza, es artificial (the Sierra) esto (the wall) es la naturaleza).'

(Lund 2006: 382)

Miguel's inversion of the normal opposition of 'natural' and 'artificial' has some rhetorical force in virtue of the way a legal act removes him from traditional life activities that draw on local materials within the park, such as the building of housing from local stone. The house is an artifice, but one that belongs to the life-history of humans dwelling in a place. To insist that persons can no longer engage in such activities itself disrupts a particular kind of causal continuity with the past.

To raise these critical comments is in no way to deny that history and the continuity of certain causal processes matter in our evaluation of environments. For reasons we develop below, and in more detail in the next chapter, there is something right about the insistence that the history and causal processes that go to the making of a place do matter. However, we will suggest that their role is more complex than Elliot claims. Our account will question whether the project of nature protection really does commit us to such policies of exclusion of human activity and the eradication of feral animals and non-indigenous plants. The approach we will adopt will suggest a much more contextual approach to issues of this kind.

On restoring the value of nature

It is restoration that presents the paradox of nature protection in its strongest form. Suppose, for example, that we try to resolve this form of the paradox by drawing a distinction between product and process. We might admit that the *process* of restoration involves various intentional human acts, but insist that this in no way prevents the *product* from being natural. However, if it is crucial to our concept of nature or wilderness that it should have originated in a certain way, we cannot ignore the question of process – of how it has come to be the way it is. Any account of a thing's origins would surely have to mention restoration if this was a process it had undergone. If a restoration had occurred at a site then it couldn't be deemed entirely natural in this sense. Hence, it seems that, strictly speaking, the idea of 'restoring nature' does involve a contradiction.

Thus it is clear that whether it is possible to restore the value of natural features depends upon what that value is supposed to consist in. If, like Katz, we think it lies in the capacity of natural entities and processes for self-realisation, then this is something that restoration may not be able to make good. On the face of it restoration is precisely not compatible with the self-realisation of nature (though Light 2000 offers several reasons for how restoration could in principle help the self-realisation of nature under Katz's description of what that would mean). As we mentioned above, Elliot, too, thinks that we cannot restore the value of natural things. According to Elliot, origin and history is not simply what makes an item natural, it is also an important constituent of its value.

Elliot makes this claim by focusing on a particular kind of pernicious restoration, namely restoration that is used as a rationalisation for the destruction of nature. According to defenders of this type of restoration, any harm done to nature by humans is ultimately repairable through restoration and therefore should be discounted. Elliot calls this view, the 'restoration thesis'. Elliot rejects the restoration thesis through an analogy between the relationship between original and replicated works of art, and nature. Just as we would not value a replication of a work of art as much as we would value the original, so we would not value a replicated bit of nature as much as we would the original thing, such as some bit of wilderness.

The force of the analogy is provided by the argument that with art, as with nature, we rely on an understanding of their origins in order to individuate them and then later to understand their value. For example, Elliot asks us to imagine a case where a developer, needing to run underground pipes through our backyard, asks to remove a valuable piece of sculpture from the yard. But because the sculpture is fragile it cannot be moved without causing it damage. The developer tells us not to worry because he will replace the sculpture with an exact replica after he finishes his work. Of course, we will reject the fake in exchange for the original because we 'value the original as an aesthetic object, as an object with a specific genesis and history' (Elliot 1995: 80). In the same way, Elliot suggests we value nature as an object with a 'special kind of continuity with the past', or with a natural genesis, understood as a 'value-multiplying or intensifying' property, though not a value in and of itself (ibid.: 81). Restoration as an attempt to reproduce nature, particularly as motivated by the restoration thesis, fakes original nature in the same way that a reproduction of a work of art fakes the original piece.

One initial objection to Elliot is that in the case of works of art, fakes are intended to deceive, whereas many if not most restoration projects have no such intention. However, his argument works just as well with the idea of a replica as it does with that of a fake and restoration does imply some attempt to replicate the original. Otherwise this would be a case of compensation. (For further objections, see Light 2000, 2002b, and forthcoming. For a much more expansive update of Elliot's view, though one which we do not believe presents problems for our objections here, see Elliot 1996.)

One other problem in Elliot's account is worth pursuing here for it will concern us further in the next chapter. We noted before that many areas that have been designated wilderness were not wilderness but the homes of indigenous people. The 'wilderness' landscapes of Yosemite were cultural landscapes both in the sense of having a meaning to those who lived there and in the sense of having been transformed by native agricultural activities. What would Elliot's response to this be? His answer would be that the valuer has made a mistake. He has misidentified the object as natural, and having discovered that this is not really the wilderness he thought it to be, he now has to revise his valuation. There is much in such a response that is right. Clearly if such a person valued the landscape because it bore no imprint of human activity a mistake has been made and valuing it for that reason is no longer appropriate. However, there remains the question as to the basis of the environmental value that such a landscape undoubtedly still has. Part of the answer to that question might involve an appeal to other values – the impressiveness of the giant Sequoia, the richness of the meadows, and so on. However, part of this assessment of value may also include the history of the place. The landscape for such an observer may now evoke something entirely new as he ponders the absence of those whose home this was and it might invoke feelings of shame or sadness. He might now look at the landscape with understanding of the interactions of human and non-human processes that went into its making, and appreciate it for that fact. The role of history, but now human history, will still play a role regardless of the causal picture determining whether something is a true wilderness. The role of history and causal processes in environmental valuation looks more complex than Elliot appears to allow. And further exploration of this complexity serves to relax the sterile grip exerted by the so-called paradox of nature restoration. For similar reasons, Eric Higgs (2003) argues that much restoration should actually be aimed at 'ecocultural' restoration rather than ecological restoration as such.

One such complexity emerges when we reflect on a form of nature restoration that does not seem paradoxical at all. It can be illustrated most simply by considering the biological analogy of regeneration. A number of organisms such as lizards, for example, have the capacity to re-grow their limbs if these become detached. Certain smaller organisms have even more remarkable regenerative capacities, including the capacity to re-grow their heads. We might call this the capacity for self-renewal. A similar capacity is observable in natural systems, and Aldo Leopold goes so far as to define the health of natural systems in terms of their capacity for self-renewal (Leopold 1949: 194), or 'natural regeneration', even though different ecosystems possess this capacity in varying degrees. Thus, some forms of vegetation are able to re-colonise a site that has been disturbed quite quickly. This suggests one distinctive sense in which restoration is possible. A system that has regenerated entirely 'naturally', that is, without the intentional or even unintentional assistance of human beings, might be said to have originated naturally, even though it no longer has the particular historical origin it once had. It has restored itself rather than been restored. Furthermore, it might on that account be said to have natural value. Notice that Elliot's analogy with works of art gives us no guidance here, and certainly cannot be used as a basis for challenging the claim, because works of art are incapable of regenerating themselves. This reflection relaxes the grip of the nature restoration paradox when we further reflect that a great deal of nature restoration consists in the facilitation of self-restoration, of our creating the conditions in which a non-intentional process of regeneration can take place. Indeed, in the most recent explication of this view (1996) Elliot does admit to degrees of restoration, preferring natural regeneration over more humanly intentional forms.

Another such complexity emerges if we consider the views of a philosopher who takes a more sanguine view of restoration – Robin Attfield:

> Only where disruption has already taken place, as . . . is the case in most of the areas of Britain recognised as 'ancient forest', is restoration or rehabilitation possible; and what is sometimes possible is reversing the damage and returning an area to a condition closely resembling its erstwhile condition in which evolutionary processes proceed independent of further human agency . . . Although Elliot's historical requirement for nature to have its full aesthetic value is not satisfied, and the area cannot be regarded as in all respects wild, there could in theory be the same blend of creatures each living in accordance with its own nature, and jointly forming a system just like the pristine one which preceded human intervention. Although the outcome is, broadly, what human agency intended, it is still equivalent to what unimpeded nature would have produced.
>
> (Attfield 1994: 49).

The reason why Attfield is more favourably disposed towards restoration is that he has a different view from Elliot as to what it is that makes natural items valuable. Attfield has an *end-state* or outcome-based view of the value of the natural world. The natural world is valuable in virtue of certain features it exhibits. Roughly, he takes the view that the natural world is good to the extent that it contains creatures living fulfilled lives. How that state of affairs is brought into existence is irrelevant to its value. If the outcome of human agency is 'equivalent to what unimpeded nature would have produced' then there is no difference in their value. Hence, restoration can be justified, if a state of affairs with the same valued features is recreated after having been temporarily disrupted. In contrast, Elliot assumes a *historical* or process-based view of the value of the natural world. It is something about the history of objects and the processes that go into their creation that make natural objects valuable.

The difference between Attfield's and Elliot's response to the problem of restoration illustrates a more general distinction between two different ways in which we value the objects and beings around us. There are some things that we value simply in virtue of their displaying some cluster of properties.

Many functional objects are most clearly of this kind. Tools, such as knives or hammers, implements like cups and plates in many everyday settings, are valued in so far as they have properties required to serve the functions they do well. If I have a hammer that is badly weighted with a loose head and someone offers me another without those faults I will take the alternative. It better displays the properties I require in a hammer.

However, many things and beings we value not merely as a cluster of properties but as particular individuals individuated by a temporal history and spatial location. We value them as spatio-temporal particulars. Our close relations to other particular persons are most clearly of this kind. While we may be able to name various virtues of those we love, their humour, kindness, generosity, wit and the like, we do not love them simply in virtue of their instantiating those properties. We value them as this particular person, our child, a sister, a friend. If in Stepford wives fashion, a replica could be created with all the properties we loved in a person plus some improvements, no one who cared properly for another would accept the replica. It is this person, with their vices as well as virtues, that we love, not simply an instantiation of a cluster of properties. Hence Montaigne's much quoted observation: 'If I were pressed to say why I love him, I feel that my only reply could be: "Because it was he, because it was I"' (Montaigne 1958: 97). Love is a relation between particular individuals as individuals. Our relations to many ordinary objects can have this significance. Thus while in general I may simply value a hammer as an object that does a job well this need not always be the case. I may value this particular hammer even with its frustratingly loose head, because this hammer was passed on to me by my father who was given it by his grandfather, who used it to make this table I sit by now, which I also value for similar reasons and despite its annoying tendency to wobble. I attach a particular significance to these objects, and that significance is a matter of their history. For that reason, things like this are said to be irreplaceable, and their loss matters in a way that the loss of other functional objects does not. The loss of my grandfather's hammer matters in a way that the loss of a hammer I bought yesterday does not. The latter I can replace, the former I cannot. It has no substitute. Still less do other persons with whom I have particular relations.

The difference between Attfield and Elliot turns in part on which mode of valuing is involved in our relations to the environments about which we care. For Attfield, what matters is that an environment contains creatures with fulfilled lives. Elliot's account does not deny that. However, his account also requires that the particular historical identity of the place matters. We think that while Elliot's account of this value needs to be developed, there is something right about his account. However, this value is not confined to landscapes that seem not to embody human activity. What Attfield's discussion highlights is the point we have made in our earlier discussion of Yosemite: history matters in the same way both in our evaluation of environments that do, and those that do not, embody human activities. Consider Attfield's example of 'ancient forest' in the British context. Most woodland in the UK is the product of human activities and will bear the imprint of human uses – coppicing, grazing, charcoal burning and so on. Now imagine a variant version of the scenario that Elliot presents, involving not 'natural' systems but humanly modified ones. Suppose that a company plans to launch an open cast mine in an ancient forest that bears the imprint of human uses, but promises restoration at the end to rectify the damage; or promises to replace the woodland with another area of woodland nearby with the same mix of trees and plants. The promise of restoration in this case would raise *exactly the same kind* of objections as those that Elliot raises for natural systems: what would be produced later would not be the same forest since its history would be different. It is the particular forest with its particular historical identity, bearing the imprint of the lives of a community that went before us, that gives the place its significance in our lives today. No modern reproduction would do. History blocks the replication of place and the substitutability of one place for another. This is as true for 'cultural landscapes' as it is for 'natural landscapes'. We aim to preserve this ancient meadowland, not a modern reproduction of an ancient meadowland because the story of the place matters. We value a spatio-temporal particular, and no reproduction will do since that particular is valued not as one possible realisation of some set of valued properties, but as a particular object constituted by a particular lineage and location. The significance of this failure of substitutability we will discuss further in chapter 11.

Restitutive ecology

To make these points about the more general role of history in our evaluation of landscape directs us to something that survives the specific problems in Elliot's analysis. A core objection to some forms of restoration ecology, such as those inspired by the Starker Leopold report cited earlier, is that they are blind to the full range of the historical significance of places. Either, like some forms of cultural restoration, they freeze places at some arbitrary point in their history, or they treat ecological sites as bundles of assets which can be maintained through reproduction. We will explore these problems further in the following chapters. However, it is certainly not the case that all forms of restoration ecology are open to the charge of blindness to history. Some acts of restoration themselves form part of the human history of our relation to the natural world and can be justified in those terms. They are not attempts to 'restore a natural state of affairs' or replicate some natural state of affairs, in the senses that Elliot and Katz outline. Nor are they simply attempts to return a place to some arbitrary 'historical' state of affairs. Rather, they receive their justification through some sense of what is the appropriate continuation of the story of a place.

Consider first an example which, from a purely ecological point of view, might seem problematic – the projects of clearing and planting that followed the destructive storms in the UK in 1987:

> In October 1987 an exceptionally fierce storm assailed south-eastern England. It challenged and sometimes transformed prevailing ideas about woods. For many people it was and still is a catastrophic intrusion which destroyed places of inexpressible beauty. Wild nature had intruded upon the ordered sensibilities of South East England. The first thought was to clear away the mess and plant new trees . . . The clearing and tree planting was part of the healing process for local communities and managers who loved the woods.
>
> (Russell 1998: 283–284)

As Russell notes, from the point of view of the recovery of woods, clearing up the mess and planting new trees was not necessarily the best response. Leaving a tangle of undergrowth and allowing the woods to recover themselves may have been a better policy in terms of the ecology of the woods. However, in terms of the relation of people to the woods that mattered to them, the clearing and planting had an expressive value. For the communities involved, it was a way of expressing care and of overcoming the loss experienced by the destruction. It may be that the care could have been expressed in more ecologically informed ways, but whatever form it took, it was as a way of continuing an historical relationship with the woods that the projects acquired their value.

Other restoration projects can express a sense of restitution for previous human wrongs. Consider the following appeals to a sense of the need to redeem past wrongs, offered in justification of the re-introduction of beavers in Denmark:

> The underlying argument seems to be that if part of nature is destroyed – in this case, if an animal species is exterminated as a result of human activity – restoration is required. This view is shared by a Danish environmental NGO called Nepenthes. A member of Nepenthes argues that restoration ecology, which admittedly differs from natural processes, can in fact help to alleviate a shared sense of moral guilt over the destruction and degradation of the natural environments:
>
> 'We say, we want this and that! It is not self-created nature, but it is exciting anyway. I find it far more constructive to go out and do something, instead of sitting back being ashamed' . . .
>
> A moral rationale for the restoration process would attach significance to the making good, or correction, of some injury – in this case, damage inflicted by us on natural ecosystems.
>
> (Gamborg and Sandøe 2004: 228)

Again the notion of recovering a 'natural state' is not at issue. Rather, restoration receives its justification within a narrative of the relations of humans to the non-human world. Restoration here is understood as a way of redeeming past wrongs, as a means of restitution.

The question remains as to whether such projects are the appropriate ways of redeeming such wrongs. Neither of the examples we introduce here may in the end be defensible as such. It may be that there are better ways of coming to terms with the loss of woodland than clearing fallen trees. And the re-introduction of beavers may not be the most appropriate response to a sense of the need to redeem past damage. However, the claim that restoration can have such restitutive value appears a quite defensible one. Ecological restoration projects need not be justified in terms of returning a place to some arbitrary natural starting point. They can play a variety of roles in the ongoing relationships that humans bear to the environments in which they find themselves.

History, narrative and environmental goods

A central claim that emerges from this chapter is that we need to take history seriously in our understanding of environmental values. One reason for taking history seriously is that the concept of the natural as opposed to the artificial is historical. The concept of 'naturalness' is a spatio-temporal concept. There is no such thing as a state or condition of something which constitutes its 'being natural', or an identifiable set of characteristics which makes any item or event 'natural'. Being natural is, and is only, determined by origin and by history: it is a spatio-temporal concept, not a descriptive one. The point is developed well by Robert Goodin. Here is the full quote from the passage that we cited earlier in this chapter:

> According to the distinctively [green theory of value] . . . what it is that makes natural resources valuable is their very naturalness. That is to say, what imparts value to them is not any physical attributes or properties that they might display. Rather, it is the history and process of their creation. What is crucial in making things valuable, on the green theory of value, is the fact they were created by natural processes rather than by artificial human ones. By focusing in this way on the history and process of its creation as the special feature of a naturally occurring property that imparts value, the green theory of value shows itself to be an instantiation of yet another pair of more general theories of value – a *process* based theory of value, on the one hand, and a *history* based theory of value, on the other . . .
>
> (Goodin 1992: 26–27)

What our remarks in the last section of this chapter have suggested however is that this historical dimension to environmental valuation is not confined to things that are natural. It applies also to humanly modified landscapes. History and process might have a more general role to play in environmental valuation. In the next chapter we will develop this narrative account of environmental values in more detail. In particular we aim to consider how one might produce a thicker account of this dimension than Elliot's notion of 'causal continuity', which we have suggested will not do. In the chapters that follow we will show how this thicker account might shed light on how we should understand two other central concepts in recent environmental discourse: 'biodiversity' and 'sustainability'. In the final chapter we consider the implications of our arguments for the nature of environmental decision making.

PART THREE

The narratives of nature

9 Nature and narrative

Thus far in this book, in criticising both broadly utilitarian approaches to environmental value and the mainstream responses to these in environmental ethics, we have defended two central claims. First, we have argued for a form of pluralism about values – one that is sceptical about attempts to force ethical reflection into the mould of scientific theories with a few ethical primitives from which our moral obligations can be derived. Those approaches impoverish the language that we bring to ethical reflection, and fail to acknowledge the real and difficult conflicts that ethical life involves. Ethical reflection needs to begin from the thick and rich ethical vocabulary we find in our everyday encounters with the environments which matter to us. Second, we have argued for the importance of history and narrative in environmental valuation. We have argued that many environmental goods we value are valued as spatio-temporal particulars. What matters are particular beings and places constituted by their particular histories.

In part three of this book we will develop some of the implications of this approach to environmental policy making. In the next two chapters we will examine how the approach might be extended to consideration of issues of policy around biodiversity and sustainability. In the final chapter we will consider its implications for understanding what makes for good decisions about the environment. In this chapter we aim to fill out our account of the role of history and narrative in environmental value. We will do so by setting out not from dramatic encounters with spectacular wild environments and the kinds of problems raised by these, but rather from encounters with some very ordinary environments that raise problems of an everyday kind. We start with a brief account of three walks in ordinary landscapes in which the role of history is clearly evident in our appreciation of their value. In those walks we encounter conservation problems of what we shall loosely call the 'old world' kind. By this we simply mean that they focus on natural landscapes that have involved a more intensive form of human habitation than is often thought to have existed in the 'new world' prior to European settlement. In the second section we offer a suggestion about how these problems should be approached so as better to respect the history and processes at work in them. In the third section we shall revisit the alternative suggestion that, in addressing problems of conservation, we should do what we can to restore a site to its original natural state. We then suggest that our preferred approach has explanatory power for a variety of conservation problems across various locales. Finally we offer an account of why history and process matter in our relationships with the natural world.

Three walks

Walk 1. The first walk was around an area called Little Langdale in the UK's Lake District National Park, and our guide was the regional manager for the National Trust, the conservation body that owns the land. Like the Adirondacks State Park in upstate New York, it is an area that is inclusive of human settlement and habitation, not a wilderness park like Yosemite which excludes such practices and habitation. Our attention was drawn to a number of problems typical of the region which bodies such as the National Trust face: how much grazing to permit; how to manage the small wooded areas; whether to fence off some of the higher slopes to allow the juniper, which still had a foothold there, to recover. Then our path turned through a farmyard and out the other side to a small mound. The mound, we learned, was a largely natural feature, slightly shaped at the edges by human hand. Recent archaeological investigation had indicated that this mound was once a 'Thingmount', or Norse meeting-place, and thus a site of some significance. Part of the mound had been excavated unwittingly by the local farmer and was now buttressed by a concrete silage clamp. So, one question was whether the Trust should aim to 'restore' the mound to its original condition by removing the clamp. The other question was raised by a more ephemeral adornment. Atop the mound and, as it were, its crowning glory, stood a thoroughly unabashed and utilitarian washing line. It was, after all, an excellent spot for drying clothes.

Question: Should the washing line be removed?

Walk 2. The second walk was around Arnside Knott, a small limestone outcrop in an area just south of the Lake District National Park known as Silverdale; this too was under the care of the National Trust. Once more we were fortunate to have the regional manager as our guide. Here, in recent memory, a certain butterfly had flourished – the High Brown Fritillary – which is relatively rare in the UK. The colony was now much reduced and there was a considerable risk that it would disappear altogether. It had flourished because of local use of the land for grazing purposes; this created the limestone grassland which the insect requires for breeding purposes. The colony began to dwindle when the practice of grazing ceased. What the National Trust has recently done is to fence off a section of the land, cut down the naturally arriving yew and silver birch which had successfully begun to recolonise the land, and reintroduce grazing.

Question: Should the butterfly colony be defended by these means?

Walk 3. The third walk was around a disused slate quarry in Northern Wales. From a landscape perspective it would normally be judged something of an eyesore – a 'scar'; and from an ecological point of view it would be judged relatively barren, showing little sign of life except for a few colonising species. For both these reasons the local authorities decided to embark on a reclamation project that would involve landscaping the area. But as work started, there was local opposition. To carry through this project would be to bury the past: it would involve burying the history of the local community and the story of their engagement with the mountain – as revealed in the slate stairways, the hewn caverns and the exposed slate face. Higher up, and most poignant of all, the workmen's huts were still in place, and inside the huts could be seen rows of decaying coats hanging above pairs of rotting boots, where the last men to work the quarry had left them.

Question: Should one let the quarry be?

History and processes as sources of value

Our objective in this chapter is not to give precise and articulated solutions to questions of the kind that we have raised, but simply to offer a proposal about how they should be addressed. The Roman Stoics said that everything has two handles, one by which it can be carried and one by which it cannot; and that one should get hold of a thing by the handle by which it can be carried (Epictetus 1910: 270). In our view, consistent with the arguments we have offered so far, one way above all not to get hold of these problems is to attempt to itemise and aggregate the 'values' of the various items which feature in the situation – the Thingmount, the washing line, the butterflies, the rotting boots – and pursue a policy of 'maximising value'. (See Holland 1995b for a further critique of the 'itemising' approach.) For besides making the implausible assumption that these values are in some way commensurable, this approach neglects their contextual nature. Second, we need to adopt a temporal perspective to understand the issues that are involved. As we argued in the last chapter, time and history matter, so that the problem can be conceived as how best to continue the narrative of the places through which we walked. From this perspective, the value in these situations that we should be seeking to uphold lies in the way that the constituent items and the places which they occupy are intertwined with and embody the life history of the community of which they form a part. This is the perspective that lies behind the following attempt to characterise (not define) the objectives of conservation: 'conservation is . . . about preserving the future *as a realisation of the potential of the past* . . . [it] is about negotiating the transition from past to future in such a way as to secure the transfer of . . . significance' (Holland

and Rawles 1994: 37). But this clearly raises a further question to be addressed, namely: What would make the most appropriate trajectory from what has gone before?

Thus far, what we are saying is that time and history must enter our environmental valuations as constraints on our future decisions. It should be observed that many of the ethical theories we have looked at previously in this volume fail in this regard, just because they do not offer a sufficiently robust place for time, narrative and history in their accounts of how we should decide on what is to be done. Utilitarianism, with its emphasis on future consequences, is a notable culprit. And while Rawlsian theory is not consequentialist, its impartial perspective (certainly in its original formulation) stretches across time, and is to that extent atemporal. From this perspective, evaluations of specific history and processes form part of that body of knowledge of particulars of which those in the original position are supposed ignorant: agents enter deliberation devoid of knowledge of the particular time and place in which they exist. Even those theories which do introduce retrospective considerations do so in the wrong way. Some deontological theories, for example, make obligation a matter of some contract that has been entered into. But in the context of conservation we are not constrained by the past in virtue of any promises we have made. The obligations we have to the past, if that is a proper way to speak here, are entirely non-voluntary. Nor are our evaluations, and the constraints on our actions, about the legitimacy of the procedures that got us where we are. Thus, for example, some historical theories of justice suggest that a distribution of goods is just if acquisitions of goods are the outcome of legitimate procedures (Nozick 1974). However, these kinds of consideration are clearly irrelevant in this context. The value a place may have, say as an ancient meadowland, generally has nothing to do with the justice or otherwise of any procedure that handed it down to us, and everything to do with the continuing historical process which it encapsulates. Similarly, the dominant theories in environmental ethics summarised in the last section of this book both duplicate some of the problems just suggested by mainstream ethical theories as well as provide little or no means by which we can evaluate the value of landscapes that consist of a mixture of human and natural environments. Nor is it only ethical theories that we find wanting in this regard. As we will argue in the following chapters, a number of the currently proposed goals of environmental policy, such as 'sustainability', 'land health', and 'the maintenance of biodiversity', fail similarly to incorporate the dimensions of time and history, and must be judged inadequate on that account.

At this point we shall no doubt disappoint some of our readers by failing to give clear criteria for what exactly constitutes an *appropriate* trajectory from what has gone before, or what the *best* way of continuing a narrative might be. The main reason is that we believe this to be a matter for reasoned debate and reflective judgement on the part of those who are involved in, or have studied, a given situation carefully and thought hard about it: it is a matter, in short, of deliberative judgement, not a matter of algorithmic calculation according to some formula that we, or others, have supplied. However, we can offer a few initial thoughts about what some of the guiding considerations might be, and hope in the process that we might pre-empt certain initial reservations to this set of suggestions.

First, the problems of nature conservation are not problems about change as such, or at least they need not be, but rather about the kinds of changes that are appropriate. Change can be too much or too little, not by any simple quantitative measure, but by a qualitative measure of degree of disruption to narrative significance. Some attempts at conservation can be disruptive precisely by virtue of stifling change and transforming the lived world into a museum piece. On the other hand, we are inclined to say that some rotting boots should be left to rot, and that some ancient monuments should continue to be decked with washing lines, rather than be removed from the intelligible temporal processes in which they feature. Some 'histories' can even intelligibly incorporate 'revolutionary' processes of fire and flood; but other dramatic forms of change are disfiguring, or worse. This is often because of their scale, pace or source. Indeed, many of our conservation problems arise out of the fact that human-induced change generally has a faster pace than ecological change, thus preventing the numerous and subtle ecological checks and balances from operating as they might. The oil slick, for example, at least for some considerable period of time, will invariably blight the potential of the marine and shoreline ecosystems that lie in its path. Cases involving so-called 'exotic' species, on the other hand, are harder to call. As we shall see, they may inhibit, but can also release, potential in terms of enhancing biodiversity, depending on circumstances.

Second, we have to acknowledge that the same site might embody quite different narratives that sometimes point to different trajectories between which we must adjudicate. The same mound of earth belongs at one and the same time to the story of an ancient meeting place, a need to dry clothes, and a farmer attempting to make a living in a world of unpredictable markets and state subsidies. There are different histories to which our decisions have to be answerable – and there are histories that, when they are unearthed, change our perceptions of the nature of a place and what it embodies. The empty hills of highland Scotland embody not just a wild beauty but also the absence of those who were driven from their homes in the clearances. Their memory must also be respected. The argument over the fate of the 100-foot-high statue to the Duke of Sutherland on Beinn Bhraggie Hill near Golspie is about which history we choose to acknowledge. (Between 1814 and 1819, the 'Black Duke', as he was known, played a leading role in evicting the indigenous people from their homes in Scotland. See Craig 1990.) We should perhaps support its removal not just on aesthetic grounds but also for what it represents to the local people, some of whom are descended from those who were driven out. For the same reasons, the now dilapidated cottages which their ancestors left behind should perhaps remain. That there is a problem about conflicting trajectories, often associated with differences of scale and pace between natural history and human history, we do not deny. But we hold that this is not a problem for our approach, but a problem revealed by our approach. It is not the task of analysis to make difficult problems appear easy, but to reveal difficult problems for what they are.

Going back to nature?

As discussed in chapter 7, an alternative approach to the conservation problems we have mentioned might be to say that, given the chance, we should do what we can to restore a given site to its 'natural state'. But, however plausible this approach might seem at first sight, we believe that the attempt to apply it in most of Europe, at any rate, and generally in sites with a more obvious history of human habitation, like the American Northeast, is beset with problems. In due course we shall see that it is also problematic in the 'wilder' parts of the US and Australia.

To begin, consider the site of our first walk, the semi-cultivated Little Langdale valley. The first question which would need to be asked here is: to which natural state should we attempt to restore it? To its Mesolithic state perhaps? Or to its natural state during the glacial, or interglacial period? To which interglacial period, exactly? And why this one rather than that? But besides the arbitrariness implicit in any of the suggested answers to this question, there is also the point that restoration in this sense would be inappropriate, in view of the ecological changes which have no doubt taken place in the mean time.

A natural response to such objections is to propose that we 'restore' the site, so far as possible, to 'what it would now be like if there had been no human intervention'. We shall not dwell on the practical difficulties lurking behind the caveat 'so far as possible', for we believe that this idea is in principle mis-conceived. More precisely, we suggest that this definition of the 'natural state' as 'how things would be, if humans were abstracted from the landscape', is radically indeterminable.

If we imagine humans removed, then we have to imagine what the situation would be like in their absence. But this, we argue, must be a radically indeter-minable state of affairs, due to the contingency of natural processes, and the arbitrary choice of starting point. Given that the slightest event may have the most far-reaching consequences – consider, for instance, the difference that may flow from the arrival or non-arrival of a particular species at a particular site at a particular time – then one can only specify what the situation would be like if all potentially relevant variables are assigned determinate values. But it would be impossible to complete such an assignment, and even if it were possible, the assignments would be bound to be arbitrary, except on the assumption of a completely deterministic universe. Second, one has to ask at what point one would begin this 'alternative' natural history, for at whatever point one chose to begin, a different history would unfold. For example, consider how we would set about specifying what Langdale would be like if there had been no humans. The absence of humans would leave niches for other species to colonise. But which species and how would they have affected the landscape? Moreover, at what point in our ancestry do we remove ourselves from the history? The fact is there is simply no saying what Langdale would be like under the supposition that there were no humans present.

Perhaps the best response to the difficulty we have just outlined is to settle for the weaker notion of what a site might be like (rather than what it would be like), save for the human presence. But this has some disadvantages. One is that it is not clear what it excludes; perhaps a site might (naturally) have come to be a pile of ashes. Another is that it becomes less clear that we have any incentive to bring about a situation just because it might have become like that naturally. This supposes that we should be prepared to endorse any possible natural world, and it is not clear that we should be so prepared.

As we suggested in chapter 7, the same difficulties arise in talking about the 'wilderness areas' in places like the US. What really is their natural state? It is not clear at all that the answer should be that these places were natural before European immigrants inhabited them. As we pointed out previously, such places were often sites of intense human habitation and cultivation, and to argue otherwise is to claim that Native Americans were somehow part of nature whereas the immigrants were not. Therefore, in these cases too, 'going back to nature' is not simply arbitrary but also ill-advised.

This position invites two criticisms. The first criticism appeals to the fact that there were large tracts of the Americas and Australasia which the Aboriginal peoples did not, in fact, penetrate. So here, at any rate, the wilderness concept remains applicable. We shall leave aside here the question of the truth or falsity of the factual claim and respond to this challenge in a different way below by arguing that our historical perspective not only provides a better account of the value of human ecological systems but of natural ones too. The second criticism appeals to a distinction between low-level human impact and 'techno-human' impact, and argues that it is only the second kind of impact which results in a seriously degraded and 'unnatural' environment. Our response to this criticism would be to insist that, even though the Aboriginal peoples may not have wrought any great devastation (though in some cases it is doubtless that they did), they will nevertheless have made a great difference to their environments.

Therefore, the conservation problem still has to be about how to continue the story in which they have been involved, rather than about how to construct a different story in which we imagine that they were never there, because their presence is somehow assimilated within the operations of the natural world.

A further reason for favouring the historical perspective rather than an approach that advocates a return to nature has to do not just with problems about the applicability of the idea of a return to nature but also about its desirability. What is so good about a nature that is characterised by minimal human influence? One possible answer to that question concerns the effect of human influence on biodiversity. Human influence, it might be argued, leads to biological and ecological impoverishment. To defend nature is to hold the line against this biological and ecological impoverishment.

To this line of argument an initial response might be: Won't the goal of achieving 'ecological health', understood as the capacity of the land for self-renewal (Leopold 1949: 194), halt ecological impoverishment just as well as the pursuit of minimal human influence? Even if wild places are reference points for ecological health, the concept is in principle as applicable to human ecological systems, such as the agricultural systems of north-eastern Europe, as to natural ones (Leopold 1949: 196). So if biotic 'impoverishment' alone is the villain of the piece, it seems that we can insure against such impoverishment by maintaining ecological health, rather than by pursuing the goal of minimal human interference.

But a more fundamental question is: What gives us the right to be speaking of biotic impoverishment in the first place? To be sure, if biotic impoverishment threatens we should no doubt do our best to avoid it; but have we earned the right to such language? As instances of biotic impoverishment, attention is often drawn to the following kinds of process: (a) habitat destruction, (b) 'pest' outbreaks, (c) increase in exotic species and (d) depletion of natural resources. But so far as the first three of these are concerned, one might ask why, exactly, are they not just 'changes in the pattern of flourishing'? So-called 'habitat destruction' might alternatively be described as a 'change in the pattern of niches'. 'Pest' outbreaks and increases in the presence of exotics, it might be said, are no more than 'changing distributions of plant and animal species'. The question which needs to be pressed is: Why are these bad changes in this situation?

The reason we press such a question is that, although we are inclined to agree that many such changes are indeed bad, we are disinclined to accept that they are always and in principle so. Consider another short walk around the region of Reposaari on the south-west coast of Finland. This walk takes one over what is in a quite straightforward sense a new land. The land is a recent post-glacial uplift from the sea – Finland's area is increasing by about 1,000 square kilometres every hundred years. The land on which Reposaari stands emerged over the last 1,000 years (Jones 1977: 15 and 59). The nearest large town, Pori, was founded as a port on the coast in 1558. The port moved with the land uplift. Pori is now an inland town, 30 kilometres from the harbour of Reposaari.

Reposaari's status as a harbour is witnessed by the quite remarkable and sometimes beautiful graffiti carved into the rock from the nineteenth century. Its history as a port is also embodied in the very particular biology of the area. The export trade from the harbour was predominantly timber, the imports generally were lighter cargoes, amongst them spices from further south in Europe. The pattern of heavy export cargoes and lighter imports meant that the ships of previous centuries arrived at Reposaari carrying ballast – soil from southern Europe – and left without it. The ballast soil deposited on the new land contained the seeds of a variety of 'exotic' species of plant which were able to flourish in the coastal climate of Finland. The result is a flora unique to the area (see Suomin 1979). The history of human activity is part of the narrative of the natural history of the area. The idea that the proper way to continue that narrative is to cleanse the area of any seed of human origin in the name of biological or ecological integrity strikes us as quite wrong, and would be properly resisted by local biologists and inhabitants of the area. The ballast flora forms as much a part of the ecology of the area as do the seeds distributed by migrating birds. The position we are urging instead is that in considering whether an 'introduced' species constitutes a harm or a good we need to consider the specific narrative we can tell about it, not whether its origins happen to be human or not.

We have one further reservation about the appeal to biotic impoverishment. Those who make such an appeal often fail to distinguish two different senses of biotic impoverishment – first as the depletion of *natural resources*, second as the loss of features of the *natural world*. The distinction is an important one because, even if it can be demonstrated that the natural world is disappearing fast, it does not follow that natural resources are dwindling. The reason is that natural resources are understood as comprising the natural world only in so far as it is capable of supplying human needs, and possibly those of other selected species. It is well recognised that human-made capital, and technology in particular, can enhance the value of the natural world in this sense, thus enabling resources to be maintained even while the natural world is dwindling. Indeed, in some cases, the increase of natural resources might require the depletion of the natural world. What often fuels concern about biotic impoverishment is concern for natural resources rather than for the natural world as such. For example, however natural hurricanes and locusts might be, they are more often conceived as a curse than as a blessing. To the extent that concern about biotic impoverishment is about natural resources, the claim that human impact necessarily contributes to biotic impoverishment has not been made.

Old worlds and new

In this section we suggest that the historical perspective we are proposing does not only have explanatory power with respect to 'old world' problems, or conservation problems involving areas with a more obvious history of human settlement and activity. It also has explanatory power with respect to 'new world' conservation problems involving less settled areas, in the sense that it does a great deal to explain the claim that we feel the natural world has upon us and wherein its value lies. For, the natural world, just as much as human culture, has a particular history that is part of our history and part of our context, both explaining and giving significance to our lives. Thus, what it is that we value about an ancient human habitation has much more in common with what it is that we value about the natural world than the new world accounts of such value would allow. The value of the natural world should be measured not in terms of its degree of freedom from human impact, but in terms of a continuity that is true to the historical processes of natural selection that it embodies. Contrary to what might be thought, our account carries value just as far out into the natural world as do the new world accounts we are challenging.

We suggest two areas in particular where the virtues of our account show through. The first is in diagnosing cases of conflict and the second is in diagnosing the tragedy of environmental loss. Taking conflict first, if we return to the conservation problems with which we started this chapter, we see that part of the tension is between human history and natural history arising from the different paces of change in the two. And it is the history of the natural world and the objects it contains that matter just as much as do those from human history; for the 'natural', too, has its own narrative dimension, its own 'natural history'. It is the fact of their embodying a particular history that blocks the substitutability of natural objects by human equivalents, rather than, for example, the inability to replicate their function. While 'natural resources' may be substituted for one another and by human equivalents – they have value in virtue of what they do for us – natural objects have value for what they are, and specifically for the particular history they embody, regardless of the scheme of value (anthropocentric or nonanthropocentric) that we use to describe that value. 'Naturalness' is a question of both origin and history; and conservation problems are frequently associated with conflicting historical narratives.

Our second illustration concerns the 'tragic' dimension of environmental loss. We do not here challenge the appropriateness of applying terms such as 'impoverishment' and 'loss' to certain environmental situations, but we do question whether reference to the ecological characteristics of these situations is alone adequate to convey the gravity of such applications. We find a hiatus here. If you look in the kitchen and find that you are running out of sugar, you might speak of this as a tragedy; but in doing so you would be conscious of your exaggeration. There is a shortfall between saying of a situation that it is unsustainable, and saying that it is tragic. The ecological characterisation fails, in our view, to capture what is at stake – fails to capture the element of tragedy which environmentalists feel. A sense of tragedy requires that there be a story – and the story can be fiction or, as Colin Macleod argues in his essay on 'Thucydides and Tragedy' (1983), fact. The historian Thucydides, Macleod writes, 'can certainly be said to have constructed his history and interpreted events, in a strict sense of the term, tragically. This is not at all contrary to his aims as a historian. History is something lived through.' We agree. It is only in the context of a history of nature, we submit, that the sense of something akin to an environmental tragedy can find adequate expression.

Narrative and nature

In the last chapter we suggested that writers like Elliot were right to suggest that history matters in our evaluation of environments. The concept of 'naturalness' itself is a spatio-temporal concept: what makes something natural is a matter of its origins and history, not a matter of some set of properties that it currently displays. However, we suggested that their role is more complex than Elliot maintains, and in particular that the concept of 'causal continuity' of present states with those of the past is too thin to capture the kind of value involved. In this chapter we have employed the concept of narrative in an attempt to thicken up the kind of account we will require to do justice to the role that history plays in environmental valuation.

However, to make these points is still to raise a series of normative questions. Why should the different historical processes that go to make up particular states of affairs or entities matter if they yield the same physical attributes in the end? What is the pull towards a historical rather than end-state evaluation in these cases? We will explore this question in more detail in the following chapters of this book. But at least part of the answer has to do with the significance of different environments in our human lives. We make sense of our lives by placing them in a larger narrative context, of what happens before us and what comes after. Environments matter because they embody that larger context. One advantage of starting from 'old world' environmental problems, rather than those in the new world, is that the role of environments in providing a larger context for our lives is here more clearly apparent. Particular places matter to both individuals and communities in virtue of embodying their history and cultural identities. Similar points apply to the specifically natural world (cf. Goodin 1992: ch. 2).

As we noted in chapter 1, natural environments have histories that stretch out before humans emerged and they have a future that will continue beyond the disappearance of the human species. Those histories form a larger context for our human lives. However, it is not just this larger historical context that matters in our valuation of the environments in which we live, but also the backdrop of natural processes against which human life is lived. Intentional human activities, such as the design of garden beds and agricultural endeavour more generally, take place in the context of natural processes that proceed regardless of human intentions and indeed which often thwart them. Habitats may be the result of particular patterns of human activity, but were not the aim of those activities: skylarks and poppies, after all, are the unintended beneficiaries of farming practices, not their direct purpose. Even in the most urban of environments one can appreciate objects for the natural processes that went into their making. Owls carefully introduced into a place have their value. But it is not the same as an environment where an owl alights spontaneously without conscious invitation into an abandoned gasworks (Mabey 1980: 140). Likewise consider the pleasure in the chance appearance of unexpected flora in a disused urban wasteland. What is often valued in particular environments is the fortuitous interplay of human and natural histories. Hence Mill's proper repulsion towards 'contemplating the world with nothing left to the spontaneous activity of nature'. In such a world we would lose part of the context that makes sense of our lives. In chapter 11 we will explore the relation of these points to some of the issues in ethical theory we discussed in the first part of this book.

However, we first want to consider the implications of our approach for some specific areas of environmental policy making. In the next two chapters we examine two concepts that have become central to environmental policy: biodiversity and sustainability. We will suggest that both have been overtaken by an ahistorical and decontextualised itemising approach that cannot do justice to the ways that environments properly matter for us. In its place we will show how the historical account we have outlined in this chapter and the last do more justice to the kinds of concern that appeals to biodiversity and sustainability are attempting to capture.

10 Biodiversity: biology as biography

'Biodiversity' along with 'sustainability' has become one of the key terms in the politics of the environment. The concept appears with increasing frequency in national and international policy documents and international declarations concerning regulation of the environment. At the international level it was one of the main items on the agenda of the UN Conference on Environment and Development (UNCED) – the so-called 'Earth Summit' – held at Rio de Janeiro in June 1992. The signing of the Biodiversity Convention at the Rio Earth Summit, which came into force on 29 December 1993, brought in its train a variety of national and local policy documents on biodiversity. Typical is *Biodiversity: The UK Action Plan*, which sets out the British Government's overall goal for biodiversity in the following terms: 'To conserve and enhance biological diversity within the UK and to contribute to the conservation of global biodiversity through all appropriate mechanisms'.

One of the attractions of the concept of biodiversity in environmental policy making is that it has the authority of science behind it. Biologists and ecologists are given the task of measuring biodiversity and developing plans to prevent its loss. At the same time concerns over the loss of biodiversity can register more widely held environmental concerns about the increasing destruction of particular habitats such as rainforests, and the threats to, and actual loss of, many biological species. At a more everyday level it can register people's concerns about changes in their familiar worlds and landscapes. Rachel Carson's influential book *Silent Spring* captures in its very title one expression of such concerns. In the UK, the disappearance of familiar features of the landscape – the red squirrel, the dragonfly and the local pond in which it was found, the skylark and the sound of its song, the local copse, the flower meadows, the hedgerow – all matter to people's everyday environmental concerns.

Appeal to biodiversity is also the focus of a variety of concerns about the actual or potentially adverse consequences of environmental changes for human welfare. Among the effects of biodiversity loss are changes in ecosystem functions, such as changes in the various 'energy cycles', changes in the viability of the food web, and the potential for soil impoverishment. Each of these kinds of systematic change can have both indirect and direct effects on human welfare. In terms of energy, take the 'carbon cycle', for example. The destruction of forests and other vegetation which absorb carbon dioxide from the atmosphere, and therefore function as 'carbon sinks', leaves increasing amounts of this gas in the atmosphere, thus contributing to global warming. This in turn will adversely affect human environments by causing changes in weather patterns, changes in the functioning of agricultural areas and changes in sea levels near coastal communities. As far as food webs are concerned, the removal of an organism which holds a key position in a food web may have effects throughout the system. Single-celled algae, for example, hold a key position in the aquatic food web, and their damage by ultra-violet radiation (due to ozone depletion) will have repercussions throughout aquatic ecosystems. So, too, with soil impoverishment. Such impoverishment can take a number of different forms, such as declining fertility, when cycles of decomposition are interrupted, desertification, when the moisture-retaining properties of vegetation are removed, and soil erosion, following from deforestation and desertification. All of these changes can have direct effects on human welfare, such as reducing the sources of food, medicines and raw materials available to us and the loss of 'models' for technological solutions to these problems that could have been found in the natural world. In addition, these changes can have indirect effects on human welfare, for example through the loss of natural crop pollinators (such as insects), and the loss of wild species which function as a potential source of new varieties for cultivation, and for the regeneration of existing cultivated varieties.

In this last context the concept of biodiversity also has appeal in a narrower economic sense when it is used to describe a resource to be exploited in global markets. Consider for example the way that economists are sometimes asked to quantify the economic value of biodiversity; or again, the way in which genetic biodiversity, both wild and agricultural, has become a resource in consequence of developments within biotechnology. Companies involved in genetic engineering, particularly those with pharmaceutical and agricultural interests, have attempted to gain property rights over particular genetic resources. They have met resistance in attempting to do so. Such developments have clear distributional implications of the kind that we have raised in chapter 4, since those genetic resources themselves are often the result of peasant and indigenous agricultural activities. Here, the question of maintenance of biodiversity is not simply one involving the existence of species as such, but extends to the histori-cal relationship between particular human communities and particular species, which is rapidly changing.

The concept of biodiversity is used then to capture a variety of environmental concerns. Indeed the way that the concept is used to capture so many different kinds of concern may give one pause to consider how far its force in many contexts is primarily rhetorical. Like the concept of sustainability which we consider in the next chapter, there are quite proper questions about whether its use in so many public debates really brings clarity to those debates. It may be that there are quite different conceptions of biodiversity that are being appealed to in such discussions. In this chapter we attempt to bring at least some clarification to the concept of 'biodiversity'. We will do so in the context of the central arguments of the last two chapters. We note first that much discussion of environmental value has tended towards what we called an itemising approach to value, which is a popular starting point for discussions about the value of biodiversity. Given the nature of biodiversity this is on the face of it somewhat surprising, and we therefore consider some of the reasons why the approach has proved popular in policy making. We then contrast this approach to the narrative approach that we outlined in the last two chapters which insists on the significance of the historical context in which biodiversity is valued. We consider the implications of this narrative-based approach for the nature of environmental policy making in regard to biodiversity.

The itemising approach to environmental values

Many discussions of biodiversity illustrate with clarity a particular approach to environmental values, which we shall call here the itemising approach (in a similar vein, Bryan Norton (2005) calls this the 'chunk and count' approach). It is an approach that dominates much ecological and economic thinking about the environment. A list of goods is offered that correspond to different valued features of our environment, and increasing value is taken to be a question of maximising one's score on different items on the list or at least of meeting some satisfactory score on each. We are conceived as having something like a score-card, with valued kinds of objects and properties, valued goods and a score for the significance of each. Policy making is then understood as the attempt to maintain, and where possible increase the total score, the total amount of value thus conceived. The approach is an illustration of a form of the consequentialism that we discussed in part one: we assess which action or policy is best in a given context solely by its consequences, by the total amount of value it produces.

The nature of biodiversity – conceptual clarifications

On one level this itemising approach to biodiversity might seem surprising. 'Biodiversity' is a shortened form of the phrase 'biological diversity'. The basic concept here is that of 'diversity'. Diversity is not itself a discrete item in the world, as the itemising approach might seem to require, but appears rather to be a property of the relations between several items. It refers specifically to the existence of significant differences between items in the world. Moreover, there are different kinds of difference. A basic distinction here is that between *numerical* difference (difference in number), and *qualitative* difference (or, difference in kind). An example of (mere) numerical difference would be the difference between two peas in a pod; we say 'as like as two peas in a pod' when we want to draw attention to how similar two things are, even though there are two of them numerically. A pea and a soy bean in contrast are different kinds of entities. While diversity refers to the existence of differences in the world, there is a conceptual contrast to be drawn between diversity and difference. Difference is usually a relation between two items; diversity, on the other hand, is usually a relation which holds amongst several items. Think for example of cultural diversity. Diversity is very close in meaning to 'variety'.

'Diversity' is a complex, multi-dimensional concept that embraces different kinds of difference – difference in weight, size, colour, etc. Moreover, it can refer to both actual and potential differences; for example, many organisms from different species are, at the embryonic stage of development, virtually indistinguishable from each other. Nonetheless, such entities are diverse by virtue of the very different adult organisms which they will become. In virtue of being a multi-dimensional and complex concept, there is no reason to assume there is any simple single measure of diversity as there is say of a uni-dimensional concept such as length or weight. This is also the case as we shall now show with the concept of 'biodiversity'.

'Biodiversity' , as we just mentioned, is shorthand for 'biological diversity' and refers specifically to diversity amongst life forms. The term attempts to capture the existence of actual and potential differences between biological entities. As we have just noted, there can exist a variety of kinds of difference. Among these are the following:

1. numerical diversity, e.g. the number of species;
2. dimensional diversity – the degree of separation, or distinction, along a dimension, e.g. difference in size, length, height and so forth;
3. material diversity – difference in the substance(s) and structural properties of which things are composed, e.g. a jelly-fish and a crab;
4. relational diversity, e.g. differences in the kinds of interactions that obtain between organisms, such as those between predator and prey, parasite and host; and
5. causal diversity – differences in the way in which things have come into existence, e.g. salmon and lungfish are quite similar but have very different evolutionary origins.

There are also different levels of difference. It is standard for example to distinguish genetic, species, ecosystem and habitat as distinct levels of diversity. Thus there is genetic diversity – the difference that obtains between organisms at the genetic level; species diversity, the most commonly recognised form of biodiversity, which underlies the prevailing focus on 'species richness' or, simply, the number of different species; ecosystem diversity, a complex measure of the extent to which ecosystems differ from one another; and habitat diversity, a complex measure of variation between habitats. Because there are these different levels of biodiversity, it is important to notice that increases of diversity at one level do not always correlate with increases at other levels.

Even within these levels there are further distinctions that it is important to register. One of these is the distinction between 'within-habitat' diversity and 'cross-habitat' diversity as described by Norton:

> Insofar as an ecosystem can be viewed as a separate and (partially) closed system occupying a delimited space, it is possible to measure its *within-habitat* diversity. This type of diversity could be designated in a rough-and-ready way by simple species counts, although many ecologists prefer measures that also indicate relative abundance of species. A system composed of one dominant species and *n* rare species is considered less diverse than a system composed of the *n* + 1 species with more even distributions. Within-habitat diversity also depends on the degree of difference among the species existing in the system. Two species from the same family add less to diversity than do two unrelated species. Diversity also has a *cross-habitat* dimension. An area is considered more diverse if it contains a number of very different systems; cross-habitat diversity indicates the heterogeneity existing among the various habitats and ecosystems in an area.
>
> (Norton 1987: 31–32)

Accordingly, suppose that two habitats contain very much the same mix of plants and animals. Then, even though they may be internally diverse, there will not be much difference between them. Suppose, next, that one of these habitats suffers a considerable loss of species. This will mean a considerable loss of 'within-habitat' diversity. But the loss might actually increase 'cross-habitat' diversity, both because there will be a greater difference between the two habitats in their mix of species, but also because there is therefore a greater possibility of their diverging from one another in the future.

Finally, biodiversity is often used as a concept that refers to the potentials of environments rather than their state at any point of time: to maintain biodiversity is to maintain the capacity of a system to issue in diversity rather than its actual diversity at a given point of time. Diversity is often not a 'manifest' property. Whereas loss of species, for example, is often a directly detectable phenomenon, changes in the diversity of complex systems can at best only be detected by indirect means. Changes in diversity may only show themselves under certain conditions, which may or may not be realised. Hence, even when we know what those conditions might be, ascertaining changes in biodiversity involves making many complex and difficult hypothetical judgements.

The attractions of itemisation

Given the complex and multi-dimensional nature of the concept of biodiversity, it looks an unlikely candidate for the itemising approach to environmental choice that nevertheless seems to dominate current policy making: when it is made operational, biodiversity is invariably approached in terms of itemisation. Thus the term 'biodiversity' is often used in practice as more or less equivalent to 'number of species' or 'species-richness'. There are good reasons for this because diversity of species will often bring with it other associated kinds of biodiversity such as those outlined above. If there are more species, for example, then there are likely to be more kinds of relations between them, and greater manifestations of material differences among them as well. So, while species loss is not identical with biodiversity loss, it is an important indicator of biodiversity loss. Hence, 'red lists' of endangered species serve a function of indicating significant losses of species that we ought to worry about. However, it should be stressed that such lists do not tell the whole story. Sometimes an increase in the population of an endangered species might coexist with a decrease in other kinds of biodiversity, and vice-versa. This can create difficult policy decisions. As was mentioned in an example cited in chapter 5, in the UK, both goshawks and red squirrels thrive in conifer plantations which are low in the diversity of their vegetation. A policy of increasing woodland diversity, which might be thought desirable on a number of grounds, would be inimical to the flourishing of these two particular species. The point is clearly an important one in the context of a policy aimed at 'enhancing' biodiversity.

Even where the complex and multi-dimensional nature of the concept is recognised, biodiversity is often operationalised in terms of the number and relative frequency of items across certain dimensions. Consider in the context of environmental policy making the following definition of biodiversity formulated by the United States Government Office of Technology Assessment:

> Biodiversity refers to the variety and variability among living organisms and the ecological complexes in which they occur. Diversity can be defined as the number of different items and their relative frequency. For biodiversity these items are organised at many levels, ranging from complete ecosystems to the chemical structures that are the molecular basis of heredity. Thus, the term encompasses different eco-systems, species, genes, and their relative abundance.
>
> (US Office of Technology Assessment 1987)

While the definition captures many of the standard observations about biodiversity, it encourages a focus on the number and relative frequency of different items: ecosystems, species and genes. Its effect, when it is operationalised, is to present us with an itemisation of species and habitat along with the injunction to maintain or enhance the numbers on the list: hence, red-lists of endangered species, or lists of threatened habitat types.

Another of the apparent attractions of this itemising approach for some policy makers is that it lends itself to economic valuations of the significance of biodiversity loss. Economic valuation of environmental goods requires defined commodities. As Vatn and Bromley note: 'A precise valuation demands a precisely demarcated object. The essence of commodities is that conceptual and definitional boundaries can be drawn around them and property rights can then be attached – or imagined' (Vatn and Bromley 1994: 137). Thus by operationalising the concept of biodiversity in terms of a list, one can arrive at a surrogate that can be commodified. Like the supermarket list, one can ask how much individuals are willing to pay for each item, were there a market, in order to arrive at a measure of the full economic significance of losses. One ends up approaching biodiversity with a list of the kind which Pearce and Moran present for the 'non-use' or 'existence' value of different species and habitats:

Species		Preference valuations (U.S. 1990 $ pa per person)
Norway:	brown bear, wolf and wolverine	15.0
USA:	bald eagle	12.4
	emerald shiner	4.5
	grizzly bear	18.5
	bighorn sheep	8.6
	whooping crane	1.2
	blue whale	9.3
	bottlenose dolphin	7.0
	California sea otter	8.1
	Northern elephant seal	8.1
	humpback whales	40–48 (without information)
		49–64 (with information)
Habitat		
USA:	Grand Canyon (visibility)	27.0
	Colorado wilderness	9.3–21.2
Australia:	Nadgee Nature Reserve NSW	28.1
	Kakadu Conservation	40.0 (minor damage)
	Zone, NT	93.0 (major damage)
UK:	nature reserves	40.0 ('experts' only)
Norway:	conservation of rivers	59.0–107.0

(Pearce and Moran 1995: 40)

The trolley full of itemised goods offers the policy maker a set of economic valuations that he or she can use in negotiations in which it is assumed that environmental goods can be traded off with other social and economic goods:

> While we cannot say that similar kinds of expressed values will arise for protection of biodiversity in other countries, even a benchmark figure of, say, $10 pa per person for the rich countries of Europe and North America would produce a fund of $4 billion pa.
>
> (Pearce and Moran 1995: 39–40)

The perceived need for itemisation is also clearly coming to the fore in the growing attempts that are being made by the biotechnology industry to commodify genetic diversity. The shift in property rights systems that allows for the definition of property rights over genetic material has seen increased activity in the collection of seeds for *ex situ* gene banks. Such banks are collections of material abstracted from their ecological setting. They are in one sense a storehouse for biological diversity and are presented as such. However, there are two problems with presenting such banks as repositories of biodiversity. The first is that they focus on genetic diversity to the exclusion of other levels of biodiversity. The second is that they focus on *ex situ* at the expense of *in situ* diversity. The maintenance of *ex situ* diversity is quite consistent with the loss of *in situ* diversity, in particular with the erosion of the particular natural habitats or local agricultural systems from which material is extracted. Diversity is reduced to a bank of discrete items which are abstracted from the natural habitats and human communities in which they developed.

More generally, what is wrong with the itemising approach to the value of biodiversity? Before attempting to answer that question we should make clear at the outset that we are *not* objecting to itemisation as such. There is nothing wrong with producing red lists of endangered species or lists of threatened habitat types that we think should be protected. They can serve an important function, not least as ways of indicating the significant losses in biological variety that are occurring. There is a place for itemisation. However, it tells at best an incomplete story about environmental values: indeed, the problem is that it leaves the story of environmental value out altogether. A purely itemising approach offers a static and ahistorical account of environmental values. The consequences are a failure to capture the significance of environmental loss, coupled with inadequate policy responses to that loss. One way to begin to see what is wrong is by considering the role that biodiversity is called upon to play in policy making under the more general goal of environmental sustainability.

Biodiversity and environmental sustainability

The concept of 'sustainability' is one which we will discuss in more detail in the next chapter. Like 'biodiversity' it has become one of the key words of the politics of the environment. Again like biodiversity, almost everyone is in favour of it. The two concepts often appear together in policy documents where the maintenance of biodiversity is presented as a central requirement for the pursuit of sustainability. English Nature (an environmental agency in the UK), for example, in their position statement on sustainable development, define 'environmental sustainability' thus: '**environmental sustainability** ... means maintaining the environment's natural qualities and characteristics and its capacity to fulfil its full range of activities, including the maintenance of biodiversity' (English Nature 1993). What is significant to note here is that in much of the economic literature sustainability is characterised in terms of each generation passing on to its successor a stock of human and natural capital. In this context biodiversity is understood as part of the natural capital that is to be maintained over time. Sustainability is taken to involve the protection both of what is called 'critical natural capital' – aspects of biodiversity which cannot be 'readily replaced' – and of a set of 'constant natural assets' – aspects of bio-diversity that can be replaced through re-creation and translocation. Thus English Nature's position statement continues as follows:

> Those aspects of native biodiversity which cannot be readily replaced, such as ancient woodlands, we call **critical natural capital**. Others, which should not be allowed, in total, to fall below minimum levels, but which could be created elsewhere within the same Natural Area, such as other types of woodland, we refer to as **constant natural assets**.
>
> (English Nature 1993)

That use of the terms 'natural capital' and 'natural assets' in this context has a strained meaning is widely acknowledged. In the context of the UK, for example, where almost all habitats have a history of human use, it is difficult to maintain the distinction between what is 'cultivated' and what is 'natural'. It becomes even more difficult when the maintenance of 'constant natural assets' is pursued through a deliberate policy of 'manmade' habitat creation or translocation. The issue becomes more complicated still when, as we saw in chapter 7, terms such as 'uncultivated' in the global context are often implicitly construed as meaning 'uncultivated by European settlers'.

What criteria for adequate 'substitution' or 'replacement' are being assumed? Capital, both natural and manmade, is conceived of as a bundle of assets. Biodiversity understood as 'capital' is broken down, in accordance with this itemising approach, into discrete components. We have a list of valued items – habitat types, woodlands, heathlands, lowland grasslands, peatlands and species assemblages. The assumption then is that we maintain our natural capital if, for any loss of one or more of these separate items, we can recreate or replace them with another of the same value. The promise of the approach is its flexibility. If a road potentially runs through some rare habitat type, say a meadowland, or an airport runway is to occupy woodland that contains some rare orchid, we can allow the development to take place *provided* we can re-create or translocate the habitat. The issue becomes one of the technical feasibility of replacement: on this turns the distinction between 'critical natural capital' (CNC) and 'constant natural assets' (CNA).

The UK Biodiversity Action Plan (HM Government 1994) states the issue of replaceability in a way that appears at first quite expansive in what it is prepared to count as critical natural capital:

> While some simple habitats, particularly those populated by mobile species which are good colonisers, have some potential for re-creation, the majority of terrestrial habitats are the result of complex events spanning many centuries which defy re-creation over decades. Therefore, the priority must be to sustain the best examples of native habitats where they have survived rather than attempting to move or re-create them elsewhere when their present location is inconvenient because of immediate development proposals.
>
> (HM Government 1994: para. 3.96. 43)

Elsewhere the plan adds:

> [H]abitats are of great significance in their own right, having developed initially through colonisation of the UK from the rest of north-west Europe after the last glaciation and then subsequently under the direction and influence of traditional human land management activities. The results of these long historical processes are not reproducible over short time scales, and indeed like individual species themselves, are a product of evolution combined with chance events which cannot be re-run the same a second time.
>
> (HM Government 1994: para. 3.95. 43)

But what is of note here is the way that time and historical processes enter the account of replaceability. The position is still consequentialist. Time and history appear as technical constraints on achieving certain results. The 'results' of 'long historical processes' cannot be reproduced readily. As another report for English Nature puts it: 'the environmental conditions that moulded them cannot be technically or financially recreated within acceptable time scales' (Gillespie and Shepherd 1995: 14). Ancient woodlands are frequently cited as particular examples of habitats 'which cannot be readily replaced'. However, the significant point is that were it possible to recreate them within acceptable time scales, then on this account it would be permissible to shift such habitats into the category of constant natural assets which are open to being reproduced.

This brings into prominence the question, 'What counts as something's being readily reproducible, or not?' In answer to the question *This Common Inheritance* (HM Government 1990) comes up with the time scale of 25 years – a human generation. We are thus presented with a picture of two contrasting kinds of habitat: on the one hand, the relatively ephemeral – the pond, secondary woodland, secondary heathland, meadowland – which can be shifted around to fit development; and on the other hand those which take longer to recreate which are therefore to be permanent features of the UK landscape. But the contrast is provisional only: the category into which any particular habitat falls becomes a purely contingent matter that is open to future revision. For the approach leaves open the possibility of technical developments in translocation or recreation skills which would eventually allow a shift in habitats from the category of those that cannot be readily replaced to the category of those that can.

This treatment of the role of time and history in biodiversity management points to part of what is wrong with the itemising approach. The root of the problem lies in its misunderstanding of the value in 'biodiversity', and of the sources of concern about its loss. If maintaining and enhancing biological diversity were simply understood as a matter of extending a list of species, habitats, and genetic material as such, then we would be constrained only by the limits of our technical creativity. We can 'create' habitats, say fenlands, for species and the tourist industry. Indeed, if it is just a list of varieties we are after, the creation of a Jurassic Park, or other developments in genetic engineering, would serve most effectively to 'enhance' biological diversity in the sense of increasing the total of species and genetic variety we have at our disposal. However, while such developments might make a contribution to the entertainment industry, it rather misses what is at stake in worries about biodiversity loss.

Time, history and biodiversity

Time and history do not enter into problems of biodiversity policy as just technical constraints on the possibility of recreating certain landscapes with certain physical properties. Rather, instances of biodiversity, such as particular habitats, are valued precisely because they embody a certain history and certain processes. As we argued in the previous two chapters, the history and processes of their creation matter, not just the physical attributes they display. The temporal processes that are taken to be merely 'technical constraints' in the UK biodiversity plan are in fact a source of the very value of the habitats and as such could not in principle be overcome. We value an ancient woodland in virtue of the history of human and natural processes that together went into making it: it embodies the work of prior human generations and the chance colonisation of species, and has value because of the processes that made it what it is. No reproduction could have the same value, because its history is wrong. In deliberation about environmental value, history and process matter and constrain our decisions as to what kind of future is appropriate. We value forests, lakes, mountains, wetlands and other habitats specifically for the history they embody. While for the very long term, notably where geological features are involved, we are often talking about histories that have no human component, for many landscapes their history includes the interplay of human use and natural processes. This is not just a feature of 'old world' conservation problems, but, as we have argued in the previous two chapters, many of the problems in the 'new worlds' are new only to their relatively more recent European settlers. Most nature conservation problems are concerned with flora and fauna that flourish in particular sites that result from specific histories of human pastoral and agricultural activity, not with sites that existed prior to human intervention. The past is evident also in the physical embodiments of the work of previous generations that form part of the landscapes of the old world: stone walls, terraces, thingmounts, old irrigation systems, and so on. And, at the local level, the past enters essentially into the value we put upon place (Clifford and King 1993): the value of specific locations is often a consequence of the way that the life of a community is embodied within it. Historical ties of community have a material dimension in both the human and natural landscapes within which a community dwells.

To take this historical dimension seriously is both more and less constraining than the approach that starts with the notion of maximising itemised biodiversity scores and maintaining stocks of natural capital. It is more constraining in that it adds to blocks on substitution. That a particular place embodies a history blocks the substitutability of one place for another. Place and nature cannot be faked, nor their *in situ* biodiversity. What matters is the story of the place. This renders even many 'ordinary' places that are technically 'easily reproducible' less open to substitution than is usually supposed. Take ponds that are a feature of landscape biodiversity that matter to people on an everyday level. Dew ponds, village ponds, local ponds more generally are relatively easy to recreate at the level of physical features such as species variety. However, the reproductions would simply not be the same places with the same meaning for a community. It is *this* pond that was used by people long ago to water their livestock, where for generations we in this local community have picnicked, fed ducks, looked for frogs and newts – *this* pond that we want to preserve. Another pond built last year could never do as a substitute simply because its history is wrong. If we want to preserve biodiversity, we need to preserve the ancient meadowland, not a modern reproduction of an ancient meadowland, not because it is difficult to reproduce, but simply because it wouldn't be an ancient meadowland. And again when we consider the sources of popular resistance to developments that change the character of a place, this is often the issue that is really at stake. The resistance is not to change per se, or at least it need not be. The sources of concern turn on the question of what kinds of changes are thought appropriate. Sometimes, indeed, delight is experienced at unmanaged untidy changes that are not intended. As we noted in the previous chapter, the unexpectedness of natural processes can be a source of value – the uninvited fauna that appear in an abandoned gasworks, the appearance of unexpected flora in a disused urban wasteland.

While the approach we recommend in some ways adds constraints, the appeal to history also relaxes inappropriate constraints on change. One unsatisfactory feature of a great deal of biodiversity management is the way that it attempts to freeze history at a certain point – insisting for example on the retention of a particular assemblage of flora and fauna. A place then ceases to have a continuing story to tell. When you look at the landscape and ask 'what happened then in the year that the managers arrived?', all there is to say is that they preserved it and nothing else happened. The object becomes a mere spectacle taken outside of history. Thus as David Russell notes in discussing the pollen diagram for Johnny's Wood in the UK's Lake District:

> [T]his is an ancient semi-natural woodland and the tendency of popular conservation is to preserve such systems in their present condition (or if it has been 'degraded' to 'restore' it to a former more natural state). This might entail measures to encourage the regeneration of the oak, for example. It is less usual, but perhaps more useful to identify and protect the processes that allow the wood to continue to function as a wood while climate and other environmental influences change as we know they will.
>
> (Russell 1998: 285)

Some biodiversity management practices, in other words, introduce inappropriate constraints on change and fail to allow the narrative of a place to continue.

Finally, appreciation of the historical dimension points to weaknesses in the very way that biodiversity itself has often been interpreted in practice. In the first section of the chapter we noted that it was surprising how the goal of biodiversity has been pursued through an atemporal itemising approach to environmental values. The historical approach we have outlined in this chapter points to some of what is lost in that treatment of biodiversity. First, and most obviously, the diversity that biodiversity embraces needs to be understood historically. Consider for example species diversity. There are strong reasons for thinking that in at least one of its basic senses the very concept of 'species' should be understood itself as referring to a historical lineage. Species are historical kinds. According to the phylogenetic theory at any rate, what makes organisms members of the same species is that they share a particular historical lineage and not that they share morphological, genetic or behavioural similarities (Ridley 1985: ch. 6; Mayr 1987; Ghiselin 1987; for critical discussions see Wilson 1999). So even if we stay at the level of species what we are concerned with is sustaining a diversity in the lineage of organisms. The point can and should be extended to the appreciation of the richness of natural and human histories embodied in the habitats and worlds into which we enter. To render them all managed in a uniform manner for the sake of 'biodiversity' would be to lose one source of richness. Moreover, it would be unlikely to realise its result. For, and this is a second aspect of biodiversity that has been lost in the itemising approach, biodiversity should be understood as a concept that refers to the potentials of environments, and not just their state at any point of time: to maintain biodiversity is to maintain the capacity of a system to diversify rather than the actual diversity manifested at any point of time (Wood 1997). In order to maintain this capacity we need to focus on the processes of change, and their variability and plurality, rather than on the static maintenance of systems at some artificially frozen point in time. Accordingly, the belief that there is some defined set of ideal management systems for biodiversity that can be globally exported and imposed on local populations is one that is likely to be inimical to the maintenance of biodiversity properly understood.

This last point applies not just to what might be called natural biodiversity, but also to agricultural biodiversity – that is, to the varieties of plants and animals developed through the history of agriculture. To sustain agricultural biodiversity also requires the maintenance of a variability and plurality of processes. Thus, for example, *ex situ* conservation of plant varieties, whatever its virtues, cannot substitute for *in situ* conservation since the processes that shape variety are lost:

> In contrast to *ex situ* conservation, *in situ* conservation permits populations of plant species to be maintained in their natural or agricultural habitat, allowing the evolutionary processes that shape the genetic diversity and adaptability of plant populations to continue to operate.
>
> (FAO 1997: 51)

The point is significant in considering the specific threats to agricultural biodiversity. The extension of property rights over genetic material so that it can be traded on markets is often seen as a solution to the problem of the erosion of genetic diversity. As we noted above it is one reason why the itemisation of genetic resources has become such a major development. The relevant object must be clearly demarcated for the purposes of the definition of property rights. However, there are good reasons for scepticism about the claim that the commercialisation of genetic resources will protect biodiversity. While the expropriation of genetic variety by the biotechnology industry relies on the existence of such variety, at the same time it threatens its continuity through the concentration of ownership and control of seed varieties on a small number of major firms. The results are likely to accelerate the loss of local variety not just in the seeds themselves, but also in the social and natural processes through which manifest biological diversity emerges. The development also has major distributional consequences through its undermining of the autonomy of local producers and by increasing their vulnerability to shifts in global markets. (For a discussion of these issues see Martinez-Alier 1997.)

So far in this chapter we have defended an historical and contextual approach to the value of biodiversity against an approach couched in terms of the itemisation of different objects of value for the purposes of maximising value. It is worth repeating here that to make this point is not to say that such itemisations have no significance. Endangered species lists can and do play an important role in indicating biodiversity loss. *Ex situ* gene banks can play an important role in the development of scientific knowledge of biodiversity. Our criticism is not of itemisation per se. What we are criticising is a particular approach to valuing biodiversity – one that takes the process of valuation to involve listing different valued features of our environment, with the aim of enabling us to maximise total value by attempting as far as possible to maximise the score for each of the different items on the list. In other words it is a particularly influential consequentialist approach to biodiversity policy to which we have been objecting. We have attempted to show why it leads to poor environmental decision making. As a corrective to that approach we have been suggesting that a historical approach to valuation offers a better basis for understanding the concerns we properly have with biodiversity loss.

The dangers of moral trumps

We suspect, however, that the approach we have advocated here will have its critics not just among those inclined to employ traditional consequentialist approaches to environmental policy but also among those who think that our approach has been too 'human centred' or 'anthropocentric'. For reasons we have already outlined in chapter 6 we do not think the distinction between 'biocentric' and 'anthropocentric' is helpful in approaching environmental policy. However, one source of concern for some environmental ethicists is that the appeal to the everyday human scale of values will not be sufficient to protect biodiversity. Its protection requires an appeal to some particular 'biocentric' or 'nonanthropocentric' moral theory that could trump the vagaries of human concerns for biodiversity conservation. Only an account of this kind would allow us to escape the dilemmas and conflicts that plural human values bring.

For example, Katz and Oechsli (1993) maintain that while there are good arguments based on human concerns to protect the Brazilian rainforest, appeals to the nonanthropocentric and non-instrumental value of the rainforest provide a more adequate justification for environmental protection. They specifically point to 'inescapable' problems that appeals to utility and justice are taken to involve. Any appeal to utility will involve uncertainty about the utility calculation of the benefits and harms from either developing or preserving an area. It would also involve incommensurable ethical claims, for example between local development and global concerns. As far as issues of justice are concerned, Katz and Oechsli maintain that an appeal by developed countries to developing countries to preserve a resource for the good of all (which would otherwise aid in the latter's development) always imposes an unjust burden on these countries. Specifically, in the case of the rainforest, Katz and Oechsli argue that first-world appeals to Brazil to forgo short-term economic gain for long-term environmental sustainability are 'imperialistic'. The need for economic development seems so great that consequences such as the effect of cutting down the forest on global warming 'appear trivial' (ibid.: 56). Instead, ascription of the value of the rainforest independent of its importance to any particular human culture should trump other human instrumental concerns and provide for a universal and impartial basis for preserving the rainforest and protecting its biodiversity. Katz and Oechsli claim that questions of the trade-offs and comparisons of human benefits, as well as questions of international justice, would then no longer 'dominate' the discussion. The nonanthropocentric value of the rainforest would trump all other considerations. And despite a closing caveat about how such an assumption is only the 'starting point' for serious discussions of environmental policy, the authors nevertheless suggest that environmentalists (not just environmental ethicists) should endorse this approach. Doing so, they claim, will enable environmentalists to 'escape the dilemmas' of utility and justice thus 'making questions of human benefit and satisfaction irrelevant' (ibid.: 58).

The idea that an appeal to some special nonanthropocentric value could in some way resolve conflicts over biodiversity by overriding the difficult issues that arise from the pull of competing and often incommensurable ethical claims appears to us to be mistaken, for reasons discussed in chapter 5. There is no reason to suppose that any special 'nonanthropocentric' justification for preserving biodiversity overrides other ethical considerations. As with ordinary moral dilemmas faced by people when they recognise competing moral claims of other humans on them, those who had recognised the valid justification of nonanthropocentric natural value would still feel the reasonable tug of competing claims of justice and human well-being. Introducing a nonanthropocentric moral theory does not produce some special set of moral trumps that overrides considerations of human justice and human interests. If it did, then nonanthropocentrism would quickly degenerate into an absurd position (see Lynch and Wells 1998, and Eckersley 1998 for helpful discussions).

Moreover, Katz and Oechsli oddly assume that the imposition of a nonanthropocentric account of the value of the rainforest on the third world would somehow not be imperialistic. Even if one were committed to the claim that the nonanthropocentric description of natural value did articulate the only true value of nature independent of human perception or human cultural perspective, it is still the case that using that conception of value to justify a halt to Brazilian development would be an imposition on the Brazilians. If it is 'imperialistic' to force the Brazilians to accept our first-world utility calculus of the value of the forest, which puts more weight on global welfare over local development, then it must be imperialistic to impose upon them our developed version of nonanthropocentrism. After all, it is not the particular forms of justification of the moral calculus that are an imposition on Brazil, but the fact that it is our assessment of the value of the rainforest and not theirs.

Finally, Katz and Oechsli ignore the empirical evidence that it was a human scale of values that motivated resistance to development in the Amazon, rather than a nonanthropocentric moral theory. To date, one of the most successful and far-reaching movements to preserve the rainforest was initiated by Chico Mendes's Brazilian Rubber Tappers Union (working in conjunction with the indigenous Forest Peoples Alliance). Tellingly, Mendes was explicit in several public appearances (including the occasion of his naming as one of the U.N. Global 500 – an annual citation of the most significant crusaders for world environmental protection) that the rationale for his protection of the rainforest was because it was his home, in fact his place of work, and not because of some abstract ethical theory that revealed the forest's value in and of itself. In 1988 Mendes was shot by agents of forestry development in response to his activities (for a summary see Hecht 1989). Katz and Oechsli are not alone in overlooking the importance of local human concerns in forming an effective strategy for fighting development of the Amazon. Susanna Hecht has remarked that many North American environmentalists have missed the real social and economic factors involved in the destruction of these forests, which locally are more understood as issues of social justice. Thus Hecht observes: 'While Chico Mendes was certainly the best-known of the rural organisers, there are hundreds of them. And many, like him, are assassinated – not because they want to save the Amazon forests or are concerned about the greenhouse effect, but because they want to protect the resource base essential to the survival of their constituents' (cited in Cockburn 1989: 45). The motivation for producing some 'new environmental ethic' that transcends debates based on a human scale of values, with human concerns, relationships, interests, delights and cares, rests on a mistake. There is no escaping debates that appeal to a plurality of particular values. Nor would such an escape be desirable. Environmental concerns have their place amongst those values. Ethical debate needs to remain on an earthly plane.

11 Sustainability and human well-being

Sustainability: of what, for whom and why?

We noted in chapter 4 that distributive issues lie at the heart of environmental problems. There are a variety of different dimensions of inequality associated with environmental change. Environmental goods and harms are distributed unevenly across class, ethnicity and gender, within and across national boundaries, within and between generations, and between species. However, much of the discussion has centred on distributive problems concerning equity over time. As we noted in chapter 4, this focus on possible inter-generational equity in the distribution of environmental goods and harms can draw attention away from pressing problems of environmental injustice within current generations. However, there are understandable reasons for this focus, since problems of inter-generational equity are brought into greatest relief by specifically environmental change. Many of the environmental harms caused by our current practices will fall upon those who will follow us, while the benefits fall upon some (although by no means all) of those who live now. The problem of inter-generational equity has been addressed within the environmental literature primarily in terms of the concepts of 'sustainability' and 'sustainable development'. Whether this approach clarifies or muddies the discussion of inter-generational equity is a moot point and one we will consider further in this chapter.

As we said at the start of the previous chapter sustainability, like biodiversity, has become one of the key goals of environmental policy. Like biodiversity, almost everyone is in favour of it. While the concept of sustainability began its life in the context of sustainable agriculture and sustainable ecological systems, since the publication of what is known as the Brundtland Report it has been used more widely, especially in the context of sustainable economic development. The 'Brundtland' formulation, taken from the 1987 report of the World Commission on Environment and Development, chaired by Gro Harlem Brundtland, states: 'Sustainable development is development that meets the needs of the present without compromising the ability of future generations to meet their own needs' (World Commission on Environment and Development 1987). The report goes on to give an intra-generational dimension to the concept of sustainability by adding that in interpreting 'needs' here, overriding priority should be given to the essential needs of the world's poor.

Since the Brundtland Report the terms 'sustainability' and 'sustainable development' appear regularly not just in the language of environmentalists, but also in the policy documents of national and international agencies and the mission statements of companies. This very consensus has, however, been a source of understandable suspicion. This suspicion is fuelled by the fact that the term 'sustainable' is used to modify nouns like 'growth' and 'development' in ways which appear to disguise the possibility of conflicts between continued economic growth and other, environmental, goals. Some environmentalists worry that the terms are simply being used as green rhetoric for continuing economic growth as usual. Concern about the concept of sustainability is at the same time apparent amongst critics of environmentalism. For example, there are those who claim that the concept is muddled and adds nothing to traditional discussion of equity. Others claim that it can serve to place environmental goals above the pressing needs of the poor.

The very elasticity of the concept has given rise to questions about what it is supposed to mean: the sustainability of what, for whom, for how long, and why? To get a purchase on just some of the ambiguities, consider, for example, the question sometimes raised in green socialist circles – 'Is capitalism sustainable?' We do not intend even to begin to answer this question (see O'Connor 1994) but to point to problems in interpreting it. The question might mean: (i) 'Can capitalism, as a global economic system, carry on into the foreseeable future?'; or (ii) 'Can particular valuable states of affairs, relationships and objects – communities, biodiversity, ecosystems, human well-being – be sustained within capitalism?' The second question itself is open to a variety of specifications. However, taking just the two basic questions, it is clear that different combinations of response are possible. Crudely speaking, one can imagine the following replies. The optimistic economic liberal might answer 'yes' to both questions, the pessimistic liberal 'no' to the first and 'yes' to the second. The pessimistic socialist might answer 'yes' to the first question and 'no' to the second. A 'no' to both questions might be an occasion for optimism (things are bad but they won't last) or pessimism (capitalism won't survive because it destroys the conditions for human existence). Different answers might be good or bad news for different groups: the answer 'yes' to the first question might be good news for one economic class, bad for another. The concept of sustainability clearly requires specification: we need to ask questions about what we are supposed to be sustaining and for whom.

Asking those questions also brings to the fore the problem of why sustainability should be of value. Not everything that we value do we necessarily think should be 'sustained', if by that one means that it should be preserved in some particular state. There are things we value in part because they are ephemeral: the rainbow, the shaft of sunlight that breaks through clouds and illuminates a distant mountain, the evening with friends, the different stages in the growth of one's children – all may have value but only in passing. No one should want to preserve or conserve them in a static state. While we may regret their passing, to the question 'wouldn't it be great if it was like this forever?' the considered answer would be 'no'. The point is one that needs to be remembered in the context of conservation. As we noted in previous chapters, a problem with some standard approaches to nature conservation is their tendency to attempt to freeze a landscape at some particular point in its development that it might be better, albeit with some regret perhaps, to allow to pass.

Economic accounts of sustainability

Sustainability, then, raises questions about what should be sustained, for whom it should be sustained and why. In the mainstream economic literature the answers to the what, for whom, and why questions run roughly as follows:

1. What is to be sustained? A certain level of human welfare.
2. For whom is it to be sustained? Present and future generations of humans.
3. Why? Either (a) to maximise welfare over time or (b) to meet the demands of distributional justice between generations.

Such responses are in the welfarist tradition of economics that we introduced and discussed in chapters 2 through 5. As will become evident, many of the questions raised in those chapters still remain central to debates on sustainability.

According to this economic literature sustainable development is to be understood as economic and social development that maintains a certain minimum level of human welfare. David Pearce, for example, defines sustainability as follows:

> 'Sustainability' therefore implies something about maintaining the level of human well-being so that it might improve but at least never decline (or, not more than temporarily, anyway). Interpreted this way, sustainable development becomes equivalent to some requirement that well-being does not decline through time.
>
> (Pearce 1993: 48)

Sustainability thus characterised raises two central questions: first, how we should understand human well-being; second, what is required so that a certain level of human welfare be maintained over time. To the first question mainstream welfare economics offers the preference satisfaction account of well-being outlined in chapter 2. Well-being consists in the satisfaction of preferences – the stronger the preference the greater the improvement in well-being. The answer to the second question is often couched in the language of 'capital'. The maintenance of a certain level of human welfare over generations requires each generation to leave its successor a stock of capital assets no less than it receives. In other words, it requires that capital – explained also as capital wealth or productive potential – should be constant, or at any rate not decline, over time. If, however, sustainability is defined in these terms one might wonder just what work the concept is doing that could not be done without it. The point is one put forcibly by Wilfred Beckerman, who asks just what the concept adds to traditional discussions of distributional issues in welfare economics:

> Making due allowance for distributional considerations means that when we are seeking to maximise total social welfare at any point of time we will be concerned with the manner in which the total consumption of society is distributed amongst the population at the point in time in question – e.g. how equally, or justly (which may not be the same thing) it is distributed. And if we are seeking to maximise welfare over time whilst making allowance for distributional considerations we would be concerned with the distribution of consumption over time – e.g. how equally, or justly, consumption is distributed between different genera-tions. Both procedures fit easily into welfare economics. Environmentalists may not be aware of the fact that it has long been conventional to include distributional considerations into the concept of economic welfare – which is a component of total welfare – that one seeks to maximise . . .
>
> (Beckerman 1994:197)

Now, Beckerman is exaggerating when he claims that distributional issues 'fit easily into welfare economics.' As we noted in chapter 4 the maximising and efficiency-based criteria employed in utilitarian approaches to welfare economics are not straightforwardly compatible with distributional criteria of choice. However, Beckerman goes on to argue that if 'sustainability' is defined in terms of maintaining certain levels of human welfare, it is not clear what else it is adding to the traditional problems that we raised in those earlier chapters about the nature of welfare and the criteria for its proper distribution. What does the concept of sustainability add to those debates? Moreover, what has any of this got to do with specifically environmental problems?

Sustainability: weak and strong

The response of defenders of the concept of sustainability in economics to the objection that it has nothing new or useful to add to the debate on the distribution of goods over time is often to insist on the importance of particular states of the natural world for the welfare of future generations. The point is still stated in terms of 'capital', specifically in terms of the concept of 'natural capital'. A distinction is thus drawn between natural and human-made capital: human-made capital includes not just physical items such as machines, roads and buildings, but also 'human capital' such as knowledge, skills and capabilities; natural capital includes organic and inorganic resources construed in the widest possible sense to cover not just physical items but also genetic information, biodiversity, ecosystemic functions and waste assimilation capacity. The distinction between the two forms of capital is taken to generate two possible versions of the sustainability requirement, a weaker version and a stronger version:

> (1) so-called 'weak sustainability': the requirement that overall capital – the total comprising both natural and manmade capital – should not decline, or (2) so-called 'strong sustainability': the requirement that natural capital in particular should not decline.
>
> (Pearce et al. 1989: 34; for discussion, see Holland 1994)

The debate between proponents of weak and strong sustainability, respectively, turns on a difference of opinion about the extent to which 'natural capital' and 'human-made capital' can be substituted for each other. As the debate is normally presented proponents of 'weak' sustainability are taken to affirm that natural capital and human-made capital are indefinitely or even infinitely substitutable (see Jacobs 1995: 62; cf. Daly 1995). Proponents of strong sustainability, on the other hand, are taken to hold that because there are limits to which natural capital can be replaced or substituted by human-made capital, sustainability requires that we maintain the level of natural capital, at or above the level which is judged to be 'critical'. Herman Daly for example offers the following definitions: '[W]eak sustainability assumes that manmade and natural capital are basically substitutes . . . Strong sustainability assumes that manmade and natural capital are basically complements' (Daly, 1995: 49).

As it is set up by Herman Daly and others the question becomes in part an empirical matter of how far 'natural' and 'human' capital can be substituted for one another and how far they are 'complements'. In fact the differences between the positions seem to be exaggerated (Holland 1997), with both sides in the debate setting up straw opponents. Thus Daly complains that the concept of strong sustainability is misrepresented by Beckerman in his criticism of the concept:

> Beckerman's concept of strong sustainability, however, is one made up by himself in order to serve as a straw man. In the literature, weak sustainability assumes that manmade and natural capital are basically substitutes. He got that right. Strong sustainability assumes that manmade and natural capital are basically complements. Beckerman completely

missed that one. He thinks strong sustainability means that no species could ever go extinct, nor any nonrenewable resource should ever be taken from the ground, no matter how many people are starving. I have referred to that concept as 'absurdly strong sustainability' in order to dismiss it, so as to focus on the relevant issue: namely, are manmade and natural capital substitutes or complements? That is really what is at issue between strong and weak sustainability.

(Daly 1995: 49)

However, the complaint can be reversed. Accounts of weak sustainability, also, are more cautious than defenders of strong sustainability often suggest. In characterising the concept Beckerman says simply that weak sustainability 'allows for substitutability between different forms of natural capital and manmade capital, *provided that, on balance, there is no decline in welfare*' (Beckerman 1994: 195, emphasis added). Even Robert Solow, who is often taken to be one of the most notable proponents of weak sustainability, asserts that 'the world can, in effect, get along without natural resources *if* it is very easy to substitute other factors for natural resources' (Solow 1974: 11, emphasis added). Neither asserts unconditionally that natural and manmade capital are substitutable. Indeed, it is hard to find any economist who makes the claim that 'natural' and 'human' capital are 'infinitely substitutable'. The most a neo-classical economist can consistently claim is that some natural and human capital is substitutable at the margins (i.e. that small marginal losses in natural capital can be substituted by small marginal increases in human capital). It makes no claims about the total substitutability of natural and human capital (Beckerman 2000; cf. Arrow 1997: 759).

Given that neither side actually holds the position which their opponents ascribe to them, what is at stake in this debate? Part of what is at stake in the debate is an empirical question as to how far substitution is possible. However, prior to that question there are some conceptual and normative questions to be addressed. What is it to say that one thing is a substitute for another? What are the criteria for saying that something is an acceptable substitute for something else? After all, there are many objects that do in fact have substitutes, but where the substitute may be seen as less acceptable than its original. 'Artificial' limbs are 'substitutes' for their natural counterparts, for example, but how far are they acceptable substitutes? Moreover, we live in a world full of ersatz goods that are 'substitutes' for the 'real' things, from margarines and coffees through to the 'habitats' and 'natures' of theme parks. The question is not just whether 'substitutes' exist, but, even where they do, how far these 'substitutes' are acceptable. How are such questions to be answered? The answers take us back to questions about human well-being and welfare which we discussed in chapter 2.

Human well-being and substitutability

Consider the first question. What is it for one thing to be a substitute for another? To understand how that question has been answered in economics we need to distinguish two conceptions of substitutability which we might call 'technical substitutability' and 'economic substitutability'.

a. Technical substitutability: We often use the concept of a substitute with respect to some specific end or purpose. We say that margarine can be used as a substitute for lard in a recipe, or that an artificial sweetener can substitute for sugar in a coffee, or that one player can substitute for another in a football game. What we mean here is that the particular object or person will serve a similar function in achieving some end – cooking a meal, making a sweet coffee or winning a game. In these everyday uses of the concept of substitute, only very specific goods can substitute for others – a lump of lard won't substitute for sugar in a coffee or for a player in a football game. On this account one good is a substitute for another if it achieves the same end, if it does the same job. A substitute of this kind might be called a technical substitute. The criterion for acceptability for a substitute on this account is roughly that the substitute does the same job or performs the same function as the original. The concept of a technical substitute is often used in economics to describe the substitutability of goods in production.

b. Economic substitutability: The concept of technical substitutability needs to be distinguished from a distinct concept of substitutability that is assumed in neoclassical welfare economics. In the theory of consumption it employs a concept of economic substitutability which allows for a much more general substitutability of goods for each other than obtains in our previous examples. The concept of an economic substitute has been mentioned already in passing in our account of substitutability outlined above. For any particular person a good, A, is said to be a substitute for another good, B, if replacing B by A does not change the overall level of welfare of that person. A loss in one good, B, can be compensated by a gain in another good, A, in the sense that the person's level of welfare remains unchanged. Why should this allow for a wider range of substitutability? The answer is that it allows that one good can be a substitute for another not if it functions to achieve the same end, but if the end that it achieves is as good for a person's well-being as the end that would have been achieved by the other good. Two goods are substitutes for one another not in the sense that they do the same job, but rather in the sense that, as Hillel Steiner puts it, 'although they each do a different job, those two jobs are *just as good* as one another' (Steiner 1994: 171). For example, on this account a lump of lard could be a substitute for sugar in the sense that the meal the person uses the lard to make is as good for that person as the coffee which they will use the sugar to sweeten. However, to make these claims is to push the question one stage back. What is it for one alternative to be as good for a person as another? What is it for goods to be substitutable in the sense of not changing a person's overall welfare? The answer to those questions depends on what account of well-being one assumes.

Standard neoclassical economics assumes a preference-satisfaction account of well-being. As we saw in chapter 2, on this account well-being or welfare consists in the satisfaction of preferences – the stronger the preference satisfied the greater the improvement in well-being. According to this view, two alternatives, x and y, are as good as each other for a particular person if they are equally preferred – if x is at least as preferred as y and y is at least as preferred as x. The person is said to be indifferent between them. Standard economic theory goes on to make a set of assumptions about the formal structure of those preferences which have the consequence that different goods are widely substitutable for each other. The assumptions are as follows:

- Preferences are transitive, i.e. if a person prefers x to y and y to z then she prefers x to z.
- Preferences are complete: for any two bundles of goods, x and y, a person either prefers x to y or prefers y to x or is indifferent between them.
- Preferences are reflexive: every bundle of goods is at least as good as itself.
- Preferences are continuous: if one bundle of goods x is preferred to another bundle of goods y, then either bundle can be fractionally altered without changing this preference ordering.

These assumptions allow the economist to construct the indifference curves that are to be found in any basic welfare economics textbook. An indifference curve is a curve that joins all equally preferred bundles of goods. If we have two goods, X and Y, we can draw a smooth continuous curve of the kind in Figure 11.1, which joins all the points of combinations of goods X and Y which are equally preferred. The slope of the curve indicates the marginal rate of substitution between goods, that is how much of one good a person is willing to give up in order to gain an improvement in the other. The standard assumption is that as a person gets more of a particular good the curve flattens.

Note that Daly's account of strong sustainability does not require us to reject this analysis. His criticism of weak sustainability turns on the claim that some goods in production are complements. Specifically Daly defends the claim that natural and human capital are complements by noting that human capital requires inputs from natural goods. A sawmill requires wood. However, even if it is the case that human capital requires some natural capital to function, this does not generate a general case for saying that a particular level and

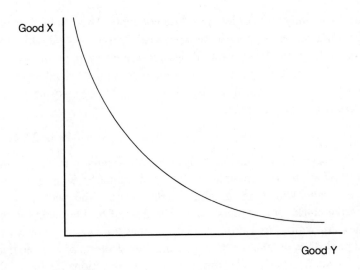

Figure 11.1 An Indifference Curve

composition of 'natural' capital as an aggregative whole is a complement to human capital similarly understood. As human capital changes with the development of new technologies and changes in human preferences, so also will the requirement for natural capital. For example, if we have substitutes for wood that are preferred to the planks from the sawmill, then the fact that sawmills require wood is something of an irrelevance. The extent of natural capital required will depend on human technology and human preferences. Imagine that humans come to prefer to live in a synthetic world of plastic buildings set amid landscapes in which, as Mill puts it, every rood of land is in fact brought into cultivation. The amount of natural capital required in such a world will be very small indeed. Hence, we do not think you can resolve debates about sustainability simply by reference to complementary goods. One needs some account of what those goods are good for, which in turn raises substantial issues about human well-being and ethical values.

Let us return to the analysis of the structure of preferences offered by neo-classical economics. Should we accept this analysis? One assumption required to make the smooth and continuous indifference curves of welfare economics is the assumption of continuity: if one bundle of goods x is preferred to another bundle of goods y, then either bundle can be fractionally altered without changing this preference ordering. Consider the following textbook counter-example to the assumption offered by Vivian Walsh:

> [T]here are some important choice situations in which the object of choice must be available in exactly one form to be *any* good at all – where the

idea of slightly more or less just does not apply. This can be so even when the thing chosen happens to be a physical thing which is highly divisible, like a chemical. Most of the drugs prescribed by physicians are highly divisible. Yet it is a matter of common practice that what he prescribes is some exact dose; quantities less that the right one may be useless and greater ones may be fatal.

(Walsh 1970: 27, cf. p.142)

Walsh's example isn't immediately a counter-example to the axiom of continuity. After all, it is possible that an agent might be willing to trade off the loss in health with a marginal gain elsewhere – the person may have an expensive crack habit to maintain and the loss of health matters little to them, or they are willing to sacrifice health for that holiday they always wanted in Blackpool. They are thus open to ignoring their doctor's advice and trading their drug for some other good. However, nothing in rationality forces that trade. There is nothing irrational about refusing any such trade-off. Indeed, if the drug is for a serious condition, to refuse any trade of the drug for some other goods might appear perfectly rational. If the condition is serious, a rational and reasonable person would follow the doctor's advice and refuse any further shift in the bundle of goods. There is no gain of welfare in any other dimension that compensates for the loss of health. The drug satisfies a need that must be satisfied if a person is to have a minimally flourishing life at all. However, the shift from preferences to needs here signals a change in the account of welfare one is assuming. Welfare is being used to refer not to a person's preferences, but to what makes for a flourishing life on some more objective account of a good life.

From preferences to needs

If one moves from a preference satisfaction to an objective state account of well-being, goods begin to look a lot less substitutable for each other. Compare for example a preference satisfaction model of well-being with one that is founded upon the concept of need. First, as we noted in chapter 2, there is a difference in the logic of the concepts of 'preference' and 'need'. The concept of need is an extensional concept, whereas the concept of a preference is an intensional one. To repeat our earlier examples from chapter 2, from 'Joseph needs glucose' and 'glucose is $C_6H_{12}O_6$', we can infer 'Joseph needs $C_6H_{12}O_6$'. However, from 'Oedipus prefers to marry Jocasta to any other woman in Thebes' and 'Jocasta is Oedipus's mother', one cannot infer 'Oedipus prefers to marry his mother to any other woman in Thebes'. Whether or not a person needs something depends on the objective condition of the person and the nature of the object – specifically, its capacities to contribute to the flourishing of a person. Whether a person prefers one object to another depends rather upon the nature of the person's beliefs about the objects. It is in part for this reason that need can be understood as part of an objectivist account of well-being. There is a corresponding difference in the way the phrase 'x is good for a' can be read, intensional and extensional. To say 'lard is good for John' could be understood as 'for John, lard is good', or in other words 'John believes that lard is good': here, 'good for' is given an intensional reading. On the other hand, 'lard is good for John' could be interpreted as 'lard improves John's conditions of life': here, 'good for' is given an extensional reading. If we move from 'good for' in the intensional sense to 'good for' in an extensional sense, we move towards an objective state account of well-being. The two readings can clearly come apart. Given that John has a weight problem, then while John's love of chips (or French fries) fried in lard might make the claim 'lard is good for John' true on the intensional reading, it will be false on the extensional reading.

A second difference between the concept of preference, as it is understood in mainstream economic theory, and the concept of need, is that the concept of non-instrumental or categorical need is a threshold concept in a way that the concept of preference is not. Non-instrumental needs are those conditions that are necessary for a flourishing life, the absence of which would be said to harm the person (Wiggins 1991). For example, a person needs a certain amount of water, food and shelter, and also certain social relations, if they are to flourish at all. A feature of such non-instrumental needs is that there are thresholds such that if a person goes below or above them her well-being will suffer. One can have too much or too little of a particular good. Recall Walsh's counter-example to the axiom of continuity. The power of the counter-example depends upon an implicit appeal to an objective account of well-being. From the perspective of the doctor making recommendations about what the patient needs to recover her health, there is a precise amount of a particular drug that the individual requires. The fact that the patient has a preference for crack cocaine or a trip to Blackpool, and is willing to trade the good, is not relevant to the appraisal of need. As far as a person's needs are concerned they are not substitutes. They do not satisfy that need and the fact that other goods satisfy some other preferences is not relevant. What if we stay among needs? The agent trades the drug for goods that meet some other need. Can't we say that the other goods are substitutes in the economic sense so long as total need satisfaction remains the same? It is not clear that we can. It is not clear that anything akin to the smooth continuous indifference curves of preference-based accounts of human welfare can be translated into the realm of need. It is certainly the case that some needs may be more pressing than others in a particular context. The need for water for a person suffering from acute dehydration might be more pressing than their need for vitamin C, and she might for that reason prioritise action to meet that need. However, where a person suffers a loss in one dimension of need that takes her below a certain minimal standard, it will not make sense to say there is a gain in some other dimension that will compensate for that loss. No compensation is possible. There is no substitute from some other dimension that will do. Hence there is no reason to assume that an axiom of continuity will hold in the domain of needs.

The points made here have more general implications for how we understand the limits of substitutability. If we move from a preference or desire satisfaction account of well-being to objective state accounts such as we introduced in chapter 2, limits to substitutability between different goods become more pronounced. Consider some standard objective list account of well-being. To live well is to have or realise particular objective states – particular forms of personal relation, physical health, autonomy, knowledge of the world, aesthetic experience, accomplishment and achievement, sensual pleasures, a well-constituted relation with the non-human world, and so on. If one allows a plurality of such goods to be constitutive of well-being, then there is no reason to assume that goods are substitutable across different dimensions of well-being. It is not the case that for a loss of good under one heading, say bodily health, there is a gain under some other, say personal relations, that leaves the person's well-being unchanged. There is, as people say in everyday parlance, no substitute for good health, for good friends, for particular places and environments. A loss in one dimension can only be properly addressed by the provision of goods in that dimension. A person who suffers from malnutrition requires specific objects of nutrition: more entertainment or better housing will not do. A person who has had no chance to develop basic capacities for literacy requires an education: offering more food will not do. To say this is not to deny that there are causal relations between different dimensions of well-being. Deficiencies in one dimension of well-being are likely to reinforce and be reinforced by those in others. (See for example Marmot (2004), on the relationship between the quality of social relations and the capacity to control one's life, and ill health.) However, remedies in the end must address deficiencies in each dimension. There are limits to the substitutability of goods across different dimensions of well-being.

Welfare is an inclusive good that incorporates a variety of different dimensions, and there is no reason to assume that a loss in one dimension can be compensated for by a gain in another. This point underpins quite rational refusals of compensation. Consider the refusal to accept monetary compensation by tribal persons facing eviction to make way for the development of the Narmada Dam discussed in chapter 4. The response is rational. A person facing eviction from the place in which the life of their community has been lived for generations, facing the disintegration of that community as their homes are flooded for a dam, can properly respond by saying that there is no good that can compensate for that loss. The loss of basic goods in the dimension of human affiliation and community cannot be compensated for by a gain in other dimensions. This is not to say that it would not be better that those who leave should receive something in compensation. However, the idea that there is a sum of money that can be offered that would maintain their level of welfare in the manner assumed by the standard economic theory is a myth founded upon a mistaken theory of welfare.

What implications do these limits have for debates around sustainability? In so far as sustainability is about sustaining or improving human welfare over generations, similar points apply. Thus understood, sustainability requires the realisation of a variety of different dimensions of flourishing that are constitutive of human well-being. What we need to pass on to future generations is a bundle of goods that can maintain welfare across the different dimensions of human life. We need to pass down the conditions for livelihood and good health, for social affiliation, for the development of capacities for practical reason, for engaging with the wider natural world. To do that requires that the goods we pass on are disaggregated. The development of capacities of reason requires formal and informal institutions for education. More and better entertainment on TV will not do. Sustaining affiliation requires sustaining the cultural and physical conditions for community, including particular environments that are constitutive of communities. An increase in the number and quality of consumer goods will be no substitute. Maintaining the capacity to appreciate the natural world and to care for other species requires us to sustain particular environments. Again, increases in entertainment will not substitute for that particular good. Different human capabilities require distinct and non-substitutable goods to realise them. It is not simply that 'natural' and 'human' 'capital' are not substitutable for each other or are in some sense complementary goods. It is rather that environmental goods are not substitutable by other goods because they answer to quite distinct dimensions of human well-being. In this respect the original Brundtland formulation of sustainability has more to be said for it than its more recent economic relatives. The Brundtland formulation is written in the language of needs, not the language of preferences: 'Sustainable development is development that meets the needs of the present without compromising the ability of future generations to meet their own needs.' Needs, for the reasons we have just emphasised, allow much less room for substitutability than do preferences. Hence the shift from a preference satisfaction account of well-being to a more objectivist account places clear limits on the substitutability of goods. In the next section we shall uncover further grounds for placing limits on the substitutability of goods.

Narrative, human well-being and sustainability

Objective state accounts of well-being are sometimes characterised as 'objective list theories' (Parfit 1984: 493). However, there are reasons to resist that formulation. To talk of an objective list is to suggest a certain view of the nature of human well-being and how well-being is to be improved. It suggests we approach well-being via a list of goods that correspond to different features of our human needs and capacities. These might typically include personal relations, physical health, autonomy, knowledge of the world, aesthetic experience, accomplishment and achievement, a well-constituted relation with the non-human world, and sensual pleasures. Increasing welfare then becomes a question of attempting to maximise one's score on different items on the list or at least of meeting some 'satisficing' score on each. While the idea of an objective list does successfully capture the existence of different dimensions of human well-being, it can lead to a failure to appreciate the role of history and narrative in appraising how well a person's life goes. The importance of temporal order is already there at the biological level. In appraising the health of some organism, the path of growth and development matters, not just some static scoring system for the capacities it has. Once we consider the cultural and social dimensions of human life this temporal order has an even stronger narrative dimension. Answering a specific question about how a person's life can be improved is never just a matter of how to optimise the score on this or that dimension of the good, but how best to continue the narrative of a life. The temporal structure of a life matters. The question is: 'given my history, or our history, what is the appropriate trajectory into the future?' Consider for example the case of a visiting rural social worker who finds an old man struggling to look after himself on an isolated farm. She suggests that he move to a retirement home. The advice of the social worker might appear perfectly sensible. It may be that from some static maximising perspective, the best course for the person would be to abandon the isolated farm that he has farmed for the last sixty years and which was farmed by his family before him, and to move to a retirement home. His material, social and cultural life might all improve. But given the way his life has been bound up with that place, the move would involve a disruption to the story of his life. The desire to stay, despite all the improvements on offer, is quite a rational one. And robbed of the physical continuity in a place that bears the imprint of the history that makes sense of his life, if he moves to the home his life might well wither away in consequence.

The narrative structure of a person's life matters to how well that life goes. Consider these further scenarios:

A. A newly married couple, couple A, go on a two-week honeymoon. The holiday begins disastrously: they each discover much in the other which they had not noticed before, and they dislike what they find. The first two days are spent in an almighty row. However, while they argue continuously over the next seven days, they begin to resolve their differences and come to a deeper appreciation of each other. Over the last five days of the holiday they are much happier and both feel that they have realized a relationship that is better than that which they had before their argument. The holiday ends happily. Sadly, on their return journey, the plane that carries them explodes and they die.

B. A newly married couple, couple B, go on honeymoon. The first twelve days proceed wonderfully. On the thirteenth day their relationship deteriorates badly as each begins to notice and dislike in the other a character trait which they had not noticed before, at the same time realizing that the other had a quite mistaken view of themselves. On the last day of the holiday they have a terrible row, and sit on opposite ends of the plane on the return journey. They both die in an explosion on the plane.

Which lives go better? Or, to stay with the language of consumer choice, given a visitation on the day before the holiday begins by an angel who presents you with a choice between the two lives, which would you choose? (The visitation and the choice itself will be instantly forgotten, so can be ignored.)

(O'Neill 1993: 53–54)

From a simple maximising perspective the final answer is holiday B: on any simple summing of goods over bads over time, holiday B contains more of the good, less of the bad. However, most individuals, given the choice, choose holiday A. They characterise the story of holiday A as a happier one than that of holiday B. What counts in favour of holiday A is the narrative order of events. Crucial to that order is the way in which the story ends. People's lives have a narrative structure, and the ending of a narrative is crucial to the genre to which a person's life, or an episode of that life, belongs – tragic, comic, pathetic, and so on. Our evaluation of how well a person's life goes depends on the narrative we can truly tell of it (O'Neill 1997b).

Similarly, for communities with their particular traditions and histories, what is taken to improve their lives from a 'maximising' atemporal perspective, may not be the most choiceworthy path to follow. Consider again the rejection of compensation by members of the local community faced with being displaced to make way for the dam in the Narmada. Even given a generous package of compensation, to shift a community in this way can involve an untold loss of the context that makes sense of its members' lives. This is not inevitable – people can prove remarkably resilient and the very process of resistance can sustain a community through a loss. But what is potentially at stake in such cases is the temporal as well as physical dislocation of a community, the loss of a context of what went before and what follows that is a condition of a meaningful life. Part of what makes for a flourishing human life is the narrative structure that gives it coherence.

This can be true even of environments where there is pressing need for improvement. For example, well-intentioned programmes of social improvement such as post-war slum clearances can fail because they treat the question of improving lives simply as a matter of improving certain conditions according to some objective list. In doing so they disregard the historical ties of community that are embodied in particular places. According to some objective list of goods, the bulldozing of slums and the dispersing of populations to modern sanitary housing might count as an improvement. However, to do this without regard for the historical ties of community and the conditions for its continuity may issue not in improvement but in social dislocation. Housing, running water, education and the like might all improve, but what can be lost are the activities and structures that make sense of it all. To say this is not of course to deny the need for change or to deny that those improvements are improvements. To reiterate a point we made earlier in chapter 8, to insist on a role for narrative and history in the evaluation of how well lives go is not to resort to traditionalism – one can make museums of communities as one can of places. Nor is it to deny that there are sometimes reasons for radical changes which are disruptive. In some circumstances, radical social change may be required for the flourishing of individuals and communities.

The role of narrative in environmental valuation that we have been examining in the last few chapters should be understood against the background of the way that narrative matters to how well the lives of individuals and communities can be said to go. We make sense of our lives by placing them in a larger narrative context, of what happens before us and what comes after. Environments matter because they embody that larger context. This is clearest in the cultural landscapes that surround us that specifically embody the lives of individuals and communities. However, as we noted at the end of chapter 8 this is true also, for reasons we have outlined in earlier chapters, with respect to natural processes. Unintentional natural processes provide part of the context in which intentional human activities take place and through which we understand their value. Hence Mill's objections to a 'world with nothing left to the spontaneous activity of nature'. In such a world, part of the context that makes sense of our lives would be lacking. Hence also the dystopian science fiction in which humans inhabit an entirely artificial world.

For reasons we have outlined in chapter 7 the role of history and narrative places further limits on the substitutability of certain environmental goods. They do not lend themselves to technical substitution for reasons we outlined in that chapter. There we distinguished between goods that we value in virtue of their displaying a particular cluster of properties, such as functional goods, and goods that we value not merely as displaying clusters of properties but as particular individuals individuated by a temporal history and spatial location. The first class of goods do lend themselves to technical substitution. If there is some other good that displays a cluster of properties that meet the same ends then that good is a substitute for the first. The second class of goods do not. There is no substitute for these goods since their value resides in their particular history. Many such goods, but not all, are also non-substitutable in the economic sense that we have been discussing in this chapter. This is most clearly true of relations to particular others. The loss of a particular family member, for example, will affect our well-being in ways that cannot be compensated for in any other dimension. However, it is true also of environmental goods. Consider for example those goods that embody particular ties of social affiliation that make for part of a good human life. For such goods there are no substitutes on other dimensions of human well-being that can compensate for their loss. Thus to return to the case we have already discussed above, faced with the flooding of a place that has been the home of some community, it is a quite proper response to say that there is nothing that could compensate for that loss. The loss is of a different kind to that involved in the loss of functional goods. It is irreplaceable in a person's life and will remain a loss for all that life might continue in its absence.

Sustainability without capital

Standard accounts of sustainability in the economic literature are couched in terms of 'capital'. In answer to the question 'what should we sustain?', defenders of weak sustainability answer that we are to sustain a certain stock of capital. Defenders of strong sustainability add the prefix 'natural' to capital: we are to sustain a certain stock of 'natural capital'. There are reasons to be wary of these answers. They use as a general metaphor for describing the relations of humans to the natural world a very specific set of relations of humans to each other and to nature that emerged in commercial societies over the last 300 years or so. It is unsurprising that it lends itself to the valuation of environmental resources in monetary terms. The term 'capital' in everyday parlance is used in a commercial context. Many proponents of strong sustainability, such as Michael Jacobs (1995), have attempted to divorce the metaphor of natural capital from this specifically commercial meaning. Jacobs takes the attempts to place a monetary valuation on 'natural capital' to undermine the distinctiveness of natural capital and hence to be the basis of weak sustainability only (Jacobs 1995: 61). However, even given that it is possible to divorce the term 'natural capital' from this specific history of the term, the metaphor is ill-suited to capture the full range of possible relations between humans and the natural world.

We noted in the first chapter that there are a number of distinct dimensions to the relation of humans to the natural world. We live from the natural world, in the natural world and with the natural world. The metaphor of natural capital is based on one way of conceiving the first of these relations, the ways in which we live from the world. As Jacobs notes, it involves construing nature in terms of its 'ability to provide humankind with the services of resource provision, waste assimilation, amenity and life support' (Jacobs 1995: 62). Now this dimension of the relations of humans to the natural world is very clearly an important one. What proponents of strong sustainability are quite right to note is that there are real physical limits to the capacities of nature to deliver those services. There are also real limits to human capacities to find substitutes for all of those natural services that are essential to meeting human needs. We have given reasons in this chapter to suggest that the assumption of ubiquitous substitutability of goods that underpins mainstream welfare economics is founded upon a questionable preference satisfaction account of well-being. However, as noted in the previous section there are also reasons to question the functional view of the natural and cultural worlds that the metaphor of capital suggests. We do not aim to sustain places simply in virtue of their providing certain services. We also value places, landscapes and objects as historical particulars that have a wider significance in human lives and to which we have a wider set of relations. We do not live in capital or stocks or bundles of assets. We live in places that have a variety of different significances for different communities and individuals. The natural world into which humans have entered and which they will one day leave is just that – a natural world with its own history; it is not 'capital'. If we are destroying marshes and forests, we are destroying places and habitats. We are not necessarily destroying 'natural capital' in the sense of assets which provide 'the services of resource provision, waste assimilation, amenity and life support'. Indeed, draining marshes and clearing forests may enhance the provision of those services. Environments, plural, are not mere bundles of resources. They are where human lives go on, places to which humans have a lived relation of struggle, wonder and dwelling. The failure of the metaphor of natural capital to capture the significance of the temporal and historical dimensions of environmental values is symptomatic of a failure to capture the different dimensions of the relations of humans to their environments.

To criticise the concept of natural capital is not necessarily to reject the concept of sustainability. The language of sustainability has power primarily because in all its technical uses it retains something of its everyday meaning where 'to sustain' is 'to maintain the life of something'. Reference to sustaining land, children, forests, future generations and communities has power since it calls upon the idea of keeping going something that has a life in either a real or metaphorical sense. For this reason the language of sustainability does have some real force and we are not claiming that the language should be abandoned. However, use of the term 'capital' is ultimately unhelpful as a way of filling out this idea. It reduces our relations to nature to one – that of resource – and loses sight of all the other relations we have to the natural world. Indeed, it is far better to hold onto the everyday sense of the term. Something that has a life has a history and potentiality to develop. To sustain the life of a community or land is not to preserve it, or to freeze it but to allow it to change and develop from a particular past into a future. Sustainability in this sense would be, to use a phrase we introduced in an earlier chapter, 'about preserving the future *as a realisation of the potential of the past* . . . [it] is about negotiating the transition from past to future in such a way as to secure the transfer of . . . significance' (Holland and Rawles 1994: 45–46).

12 Public decisions and environmental goods

What makes for good public decisions about the environment? We saw in part one that the dominant answer to that question has its basis in a particular maximising form of consequentialism. A good decision is one that best improves the well-being of affected agents. A form of consequentialism is embodied in many of the standard tools for public decision making, most notably in cost-benefit analysis. In the first part of the book we argued that this dominant answer to the question is inadequate. It fails to recognise constraints on policy making that are independent of the realisation of total well-being, constraints that have been articulated in different ways from within the traditions of deontological and virtues ethics. It is also unable to deal properly with the distributive dimensions of environmental decision making. Furthermore, the existence of plural and incommensurable values is incompatible with the assumption it requires that there is a measure of value through which we can arrive at a maximising score for the value of options. In chapter 5 we examined how rational public decision making might proceed in the absence of some single measure of value, looking in particular at procedural alternatives.

In the second part of the book we have argued that the main response to our environmental problems from within environmental ethics – the demand for a new environmental ethical theory – is misconceived. Against this mainstream approach we have suggested that consideration of environmental policy needs to proceed from the human scale of values. By this we do not mean that the only things that are of value are human beings, but rather that we need to start from our human relations to other beings and to the worlds we inhabit, and from our human responses to those beings and worlds. Those responses are open to education and change, but it is with such responses that ethical deliberation does and should begin. Much of the literature on environmental ethics has been too concerned with abstract schemes of values rather than with ethics on an earthly plane. An ethics that is rooted in the everyday is forced to recognise the conflicting plural values that make claims upon us. There is no single value to which all others are reducible, or some value that trumps others. Correspondingly, in making choices, there is no way of short-cutting difficult processes of deliberation between competing claims.

We also argued in part two that in so far as the concept of the 'natural' does play a role in our responses to the environments we inhabit it points to the significance of history and narrative in those responses. However, we suggested that this response is not confined to what is natural but reaches out also to cultural landscapes and objects, and to our relations to our fellow human beings. History and narrative matter. To make this point is not to say that it is the only thing that matters – we think that view would be straightforwardly false. But it does matter. The narrative structure of human lives matters to how well those lives can be said to go. Many of the beings and worlds that we value we do so, not as clusters of properties that satisfy some criterion or meet some particular end, but as spatio-temporal particulars, with a particular history. We argue that the value of both cultural and natural landscapes is far better understood in these terms. In the final part of this book we have attempted to develop our narrative approach in more detail through an examination of two of the concepts that are central to environmental policy. We suggest that it serves as a necessary corrective to the itemising approach that otherwise shapes and distorts biodiversity policy. And in the context of sustainability we suggest that it makes visible certain blocks on the substitutability of different goods – a substitutability that the assumptions of mainstream welfare economics otherwise render ubiquitous.

In this final chapter we return to a question that was central to our discussion in the first part of the book: what makes for good public decisions about the environment? We will first outline the main deliberative response to the standard utilitarian approach to environmental decision making, and we will outline one way of understanding the two approaches – as a contrast between substantive and procedural accounts of rationality. We will argue that while deliberative approaches to environmental decision making offer important insights into what is involved in making good public decisions, they often tend to share with consequentialist approaches a particular picture of decisions as discrete events that are to be appraised as such. We will argue that this picture of decisions is misleading. Decisions are not always discrete events, and even where they are they should not be appraised as such. They can only be properly appraised in terms of historical patterns of choices through which the character of institutions is expressed and developed. The rationality of decision making itself has a historical dimension that existing accounts ignore.

Procedural rationality and deliberative institutions

Normative accounts of what makes for good or rational decision making are dominated by two main perspectives on what counts as rational behaviour, which Herbert Simon has influentially characterised as 'substantive' and 'procedural':

- 'Behaviour is substantively rational when it is appropriate to the achievement of given goals within the limits imposed by given conditions and constraints.' (Simon 1979: 67)
- 'Behaviour is procedurally rational when it is the outcome of appropriate deliberation.' (Simon 1979: 68)

Classic consequentialist models of what makes for a good decision assume that decisions should meet norms of substantive rationality. The right decision is one that produces the best outcome, where on the standard utilitarian account the best outcome is defined in terms of the maximisation of well-being. Purely procedural accounts of what makes for a good decision are in contrast backward looking rather than consequentialist in nature. A good decision is one that is the outcome of rational deliberation. While more sophisticated indirect forms of consequentialism are consistent with procedural accounts of rational choice, pure procedural accounts are often associated with a deontological ethic. Indeed, given that Kant's categorical imperative is interpreted as a principle of rationality, Kant's own account of moral choice could be understood as procedural in form. A moral decision is one that accords with the norms of reason.

As we saw in part one of this book, the most significant practical expression of the utilitarian approach to decision making is cost-benefit analysis, which is still the main tool employed in environmental decision making. However, while this approach has remained dominant in public decision making, there has lately been renewed interest in a more deliberative approach to environmental choices which often calls upon procedural approaches to rational choice. Deliberative models of democratic decision making, in particular, have enjoyed an increasing popularity, especially in the area of the environment. Deliberative models of democracy are on the standard accounts contrasted with aggregative models of democracy. On the aggregative account, democracy is a procedure through which the preferences of citizens can be aggregated so that decisions can reflect those preferences effectively. In contrast, the deliberative theorist argues that democracy should be understood as a set of procedures through which preferences are formed and transformed through reasoned dialogue between free and equal citizens (J. Cohen 1989; Dryzek 1990, 2000; Elster 1986; Miller 1992, for discussion of the use of deliberative democracy in the environmental sphere see Smith 2003; Norton 2005; O'Neill 2007: chs. 9–12).

This deliberative approach to public institutions for decision making can be understood in terms of a procedural account of rational behaviour. What makes a choice rational is that it is the outcome of appropriate deliberation. The source of most recent work on deliberative institutions has been Kantian. Deliberative institutions embody the norms of public reason which Kant took to define the enlightenment: 'reason has no dictatorial authority; its verdict is always simply the agreement of free citizens, of whom each one must be permitted to express, without let or hindrance, his objections or even his veto' (Kant 1933: A738/B766). Kant's claim about the nature of reason has been echoed in recent defences of deliberative democracy by Jürgen Habermas. Institutions are rational to the extent that they are free from the exercise of power, from strategic action which aims at moving actors without persuading them by reason, and from self-deception by actors, such that the judgements of participants converge only under the authority of the good argument: 'no force except that of the better argument is exercised' (Habermas 1975: 108).

There have been a number of recent experiments attempting to bring this deliberative approach directly into public decision making. These include institutions such as citizens' forums, citizens' juries and consensus conferences in which a small number of citizens are brought together to deliberate on particular questions of public policy for a number of days. The forums normally allow citizens to call and question expert witnesses on the subject of their deliberations. The experiments have been particularly prevalent in the environmental sphere (Smith, 2003). Clearly, the degree to which any particular citizens' forum meets the norms of reason of the kind outlined by deliberative theorists is open to question. Citizens' capacities, and confidence to speak and to be heard, can be unevenly distributed across gender, class and ethnicity. The participants do not control the agenda for deliberation, nor who is invited to speak, and how the process opens and closes. Those features do allow the process to be potentially used strategically by powerful interests. However, while the actual practice of such initiatives in deliberation can depart from ideal norms of rational deliberation, the deliberative theorist can properly respond that the norms of rational deliberation guide behaviour and provide the standards from which it can be judged.

We have already argued in chapter 5 that this deliberative approach has real merits. We noted in that context that it offers an account of rational public decision making that is consistent with the existence of plural and incommensurable values – a situation in which there is no single measure of value that we can employ to trade off different values to arrive at a final choice. On that account decision making is not a matter of calibrating losses and gains on various values according to some measure, and then deciding which produces the highest total value. It is rather a matter of attending to different reasons and forming a judgement through the process of deliberation. We suggested in chapter 5 that a procedural account of rationality is better able to offer us an account of how rational choice is possible in the context of the plural and conflicting values we typically find in environmental choices.

To make these claims in defence of the deliberative approach is not to say that it is free from problems. Questions have been raised about how those who have no voice can be represented in deliberative fora. For example, how are the interests of future generations and non-humans to be included? Questions have also been raised about the kinds of argument and communication that the Kantian account of deliberation permits and precludes in public deliberation, and hence about who can be included in deliberation (O'Neill 2007: chs. 9–12). We will not examine these issues here. Rather, we want to consider the procedural model of rational decision making that underpins the deliberative approach. We want to suggest that while the deliberative model has been an important corrective to purely substantive accounts of rational choice and behaviour, it shares with the substantive account a particular picture of the nature of decisions and the way they should be assessed. Both tend to espouse a picture of decisions as discrete events which are to be assessed as such. We will argue the need for a more historical approach to decisions and their assessment that is concerned not just with particular decisions but with the character of public decisions that is exhibited in a pattern of choices over time.

Decisions in context

Both standard substantive and procedural accounts of rational choice tend to share a particular picture of the nature of decision making and how it should be evaluated. The picture is one of an individual or group of individuals deliberating on some choice and then arriving at a moment of decision. The rationality of the decision is to be judged either in terms of whether it is appropriate to achieve some outcome (substantive) or whether the deliberation that preceded it met norms of reason (procedural). Two features of that picture are worth noting here. The first is that decisions are understood as discrete events, with particular processes that precede them and particular consequences that follow. The second is the assumption that they should be evaluated as such, as discrete events appraised either in terms of preceding deliberation or in terms of subsequent consequences. In particular, the appraisal takes place in isolation from consideration of the prior history of choices by the decision-making agent. Both assumptions need to be questioned.

Consider first the assumption that decisions are discrete events. There are certainly cases of decision making in which we can date decisions. A committee chair looks around the room and says, 'so we are all agreed'; there are nods and murmurs of assent and the decision is minuted by the secretary. Prior to the decision there was a process of deliberation. Similar assumptions are often made about individual decisions. A person can say 'I remember the day and time that I decided to . . .' Prior to that we might imagine a struggle between difficult options before coming to the clear and dated decision. These are regarded as paradigm examples of decisions and they inform much work on decision making. However, it is easy to be misled by such examples, which on reflection turn out to be the exception rather than the rule. 'I remember the day I decided to X . . . ' is the sort of claim that appears in biographies, but it is often a story-telling artefact. In our actual lives we frequently find ourselves having made major decisions without being able to put a date on any decision. Many of the most momentous decisions we make are impossible to date. Consider for example the particular career paths we chose to follow. It would be difficult for any of us, for example, to date the moment when we decided to become philosophers or to spend our lives ministering to the sick. We frequently find ourselves having decided to do things without there being any particular discrete moment of choice. Neither, in many cases, was there a process of deliberation that specifically preceded a particular moment of decision. And in some of these cases there was no moment of decision separate from the action itself. Rather is it the case that our decisions, and actions, are often the outcomes of habitual ways of carrying on, the routines of our lives, ongoing deliberations about our lives or of unrehearsed responses to fortuitous circumstances. To the question 'and when did you decide to become X or do Y?' the honest answer is that there was no discrete moment of decision. In short, reflection on actual decision making tends to reveal that what are taken to be paradigm examples are no more than one end of a broad spectrum of cases.

Similarly, at the level of public decisions within organisations it is often difficult to find a moment of decision. In a similar way, decisions emerge not from a specific process of deliberation, but from organisational routines and habitual ways of carrying on, and sometimes from the unreflective pursuit of standard procedures. Decisions are not a matter of some particular actor choosing some action after a particular process of deliberation. Rather, as Graham Allison in *Essence of Decision* notes, organisational behaviours can be understood 'less as deliberate choices and more as the *outputs* of large organisations functioning according to standard patterns of behaviour' (Allison 1971: 67). Everyday acts, whether of firms or of governments, and that have environmental impacts, are often the outcome not of discrete decision but of organisational routines. Consequently, just as in individual cases, *when* decisions are made can be indeterminate. There is no moment of decision. On occasions the absence of a clear moment of decision is visible and has consequences. For example, a British Cabinet Minister, Stephen Byers, lost his job because of a lack of clarity about when a decision – involving Railtrack – was supposed to have occurred. What is taken to be the paradigm model of decision making therefore is often misleading as an account of how decisions are actually made. Decisions are not always discrete events. Indeed, the presentation of a decision as a discrete event can itself be an institutional artefact, a retrospective formal ratification of decisions already taken.

Now one response to these points is to say of these cases, both individual and collective, that there are no discrete decisions because no decisions were made.

None of us 'decided on' our careers – this is something that just 'happened'. The argument would run that we need to draw a strong distinction between acts that are the outcome of habits and routines, and there are many of these in personal and public life of the kind we have mentioned, and acts that are the outcome of decision. The former are unreflective acts that are the results of dispositions of character or organisational routines; the latter are reflective and intentional acts that are the outcome of deliberations which led to a moment of decision.

This response is too strong for two reasons. First, in many of the cases we described there will be a deal of deliberation that goes on without there being a moment of decision. I think at various points about the kind of life I want to lead, and reflect on how I have led it, without there being a moment of choice where I say – 'now it is decided'. The absence of a discrete moment of decision should not be confused with the absence of deliberation. Second, our habits and routines are themselves the possible objects of deliberation and decision. We can reflect upon them, make resolutions to change them, some even dated, say on New Year's Day. It is because we can reflect upon, deliberate about and change our habitual ways of proceeding that we bear responsibility for them and for the outcomes of habitual action. It is no excuse for some behaviour – say, of sloppiness that leads to an accident – simply to remark: 'I couldn't help it – sloppiness is a bad habit I've got into'. The obvious retort is that you should get out of it again. The dispositions that form the character, both of individuals and of institutions, are themselves open to deliberation and change. The point is important to consideration of the second assumption made in many models of deliberation – that we assess decisions as discrete events.

The second assumption of standard accounts of decision making is that decisions should be evaluated simply as discrete events to be appraised either in terms of preceding deliberation or ensuing consequences. This assumption also needs to be questioned. Certainly the prior processes of deliberation and subsequent consequences of particular actions often do matter to how we appraise a decision. Nothing in the following is intended to suggest otherwise. However, they are not all that matter. Even where a decision is a discrete event, the appraisal of any such decision cannot take place in isolation from the prior historical pattern of choices by an agent, individual or collective. Even where a decision is a single discrete event, it cannot be evaluated as such. The goodness of a decision can only be properly appraised in the context of a pattern of decisions over time. The historical pattern of agents' decisions and choices provides the context through which we make sense of and appraise particular decisions. Here, a virtues-based perspective that pays attention to the pattern of an agent's decisions, has something to offer as a corrective to the dominant consequentialist and deontological accounts that dominate the literature on decision making.

Aristotle's account of deliberation and choice offers a useful starting point to any such account. Deliberation is at the centre of Aristotle's account of choice making. But alongside and equally important is the role played by the habits and character whose formation is interlinked with these processes of choice and deliberation (Burnyeat 1980). Character and decision are related. On the one hand we form our character through the decisions we make. It is through particular choices that our habits and dispositions of character are formed. We become generous by performing generous acts. If someone is tight-fisted and knows it but wants to change, they change their character by performing generous acts and learning to enjoy them. On the other hand we express our character in the particular decisions we make. The same act of buying a round of drinks may in one person be an expression of generosity, in the other an act of toadying to curry favour with some particular group. To make such judgements one cannot treat the choice and act in isolation – one needs to know something of the history of a person's choices in order to appraise his or her character.

Similarly, what might be called the character of organisations and institutions is formed by particular decisions and is expressed through those decisions. It is through particular choices that the habitual behaviours that define the nature of organisations are formed and it is through choices that the nature and character of institutions are expressed. Consider some of the terms that we apply to organisations when we criticise decisions they make and actions they undertake: they can be said, for example, to be untrustworthy, impersonal, uncaring, disorganised, exploitative. In making those judgements we do not consider any decision in isolation, even where there is a decision to be isolated. It is rather the history of decisions over time that matters. The terms we use describe kinds of institutional vice. Take, for example, the trustworthiness or otherwise of institutions. One cannot make a judgement about the trustworthiness of an institution only on the basis of a particular decision made at a particular point in time. A particular decision might be an exemplar of rational choice on given models. Substantive and procedural norms might be met. However, there may be good grounds for scepticism if that decision is part of a wider pattern that exhibits institutional vices.

The recent growth in deliberative institutions itself offers a possible example of where such scepticism is due. One reason for the growth of interest in consultation and deliberation in the UK has been to address the 'deficit of trust' in public institutions. Typical symptoms of that deficit are taken to be the decline of participation in formal electoral processes, the increasingly expressed distrust in elected representatives and the rise in the use of direct action. Formal experiments in deliberative institutions are in this context often presented as attempts to 'restore trust' and hence 're-engage the public' in the political processes. One rational response to at least some of these developments is to treat them with some initial scepticism, even if a few particular decision-making procedures appear exemplary. Certain developments of this kind aim not to make institutions more trustworthy, that is, to be more deserving of trust, but rather to give them the appearance of trustworthiness that they may not deserve. To be able to express trustworthiness requires not a single individual exemplary decision-making procedure, or even a few of these. It requires a pattern of such decisions over time.

As another example, consider the trustworthiness of institutionalised decision procedures themselves. Take cost-benefit analysis which we discussed in part one. One of the most powerful criticisms that can be made of cost-benefit analysis as an institutionalised practice applied to large public projects is that it is systematically wrong. Flyvbjerg et al. (2003) note that over a remarkably long period of time, over a wide range of countries and on a variety of different kinds of project, cost-benefit analyses of public projects systematically and substantially underestimate the costs of projects and tend also to overestimate their benefits. Throughout that history there is no evidence of institutional learning. The same errors are repeated. How should one respond to such findings? One response is that for any particular cost-benefit analysis, no matter how much we are reassured by relevant experts of its soundness, we have good grounds, as citizens, for scepticism. However, such systematic discrepancies between projected and actual costs and benefits also give grounds for wider questioning of the trustworthiness of the institutions in which the procedure is embedded. For example, one possible response is that there is systematic deception at work, and deception that is not entirely surprising, given that those involved in the processes of assessing such projects often have an interest in their development (Flyvbjerg et al. 2002). However, if one is to employ the language of deception here it should not be focused on individual acts of lying, although these might sometimes occur, or on the dishonesty of individual participants in the processes of assessing public projects, although there may sometimes be dishonest politicians and bureaucrats involved. Given that the distortion of costs and benefits is systematic, it is rather a matter of institutional structures, procedures, habits and norms. If there is deception involved in these cases, it will include self-deception by the participants in the process of assessment, in addition to the wider deception of the public. It is the institutional procedures and structures of power and interest that are the proper object of assessment, rather than merely individualised decisions, acts and agents.

Consider, as a particular example, the high profile and tragic case of the sinking in 1987 of the car ferry *Herald of Free Enterprise* in which 189 people died crossing the English Channel. The immediate cause of the boat's capsizing was an inrush of water that was the result of the car ferry doors being left open. The immediate human failures leading to the event were also noted in the subsequent public inquiry. For example, the assistant bosun responsible for shutting the doors was asleep following his being relieved from previous duties. The bosun who did notice that the doors were open did nothing since shutting the doors did not form part of his duties. No one told the captain the doors were open – and indeed there were no procedures for telling the captain that this was the case. The captain did not provide clear orders. However, those immediate events themselves take place against a wider background that includes not just the unstable design of the ferry itself, but an organisational pattern of procedures and routines. Thus the Sheen report on the disaster makes the following claim:

> At first sight, the faults which led to this disaster were the aforesaid errors of omission on the part of the Master, the Chief Officer and the assistant Bosun, and also the failure by Captain Kirk to issue and enforce clear orders. But ... the underlying or cardinal faults lay higher up in the Company. The Board of Directors did not appreciate their responsibility for the safe management of their ships. They did not apply their minds to the question: What orders should be given for the safety of our ships? The directors did not have any proper comprehension of what their duties were. There appears to have been a lack of thought about the way in which the *Herald* ought to have been organised for the Dover/Zeebrugge run ... From top to bottom the body corporate was infected with the disease of sloppiness.
>
> (Report of the Court, No. 8074. Department of Transport,
1987, para.14 M.V. Herald of Free Enterprise)

The vice of sloppiness identified here is ascribed primarily to the corporate body rather than to specific individuals. The individual acts of carelessness exhibited in the immediate events that led to the disaster are but the manifestation of a more general pattern of procedures and routines. This is sometimes what people mean when they say, 'it was a disaster waiting to happen'.

Similar points apply elsewhere within the corporate sphere, for example, in the patterns of preventable industrial accidents and more pertinently here also in the patterns of environmental pollution, both major and minor (Slapper and Tombs 1999). The capsizing of oil tankers, like the capsizing of car ferries, is often the result of a pattern of procedures rather than of a particular conscious decision. Consider the accident involving the *Exxon Valdes* oil tanker which caused a major and much analysed environmental disaster off the coast of Alaska in 1989. As the report on the disaster notes, while the behaviour of the captain formed the immediate object of criminal prosecution, the causes lay in changes in habitual organisational procedures:

> Without intending to minimize the impact of [Captain] Hazelwood's actions, however, one basic conclusion of this report is that the grounding at Bligh Reef represents much more than the error of a possibly drunken skipper. It was the result of the gradual degradation of oversight and safety

practices that had been intended, 12 years before, to safeguard and back-stop the inevitable mistakes of human beings.

(Alaska Oil Spill Commission, 1990,
Spill: The Wreck of the Exxon Valdez)

Elsewhere the report refers to other institutional vices: 'success bred complacency; complacency bred neglect; neglect increased the risk until the right combination of errors led to disaster.' (Alaska Oil Spill Commission, 1990, *Spill: The Wreck of the Exxon Valdez.*)

Such patterns of institutional procedures themselves need to be understood against the background of wider institutional structures within which organisations operate, for example the institution of the market where there is intense competition for cutting costs. Getting the proper level of focus on the character of public institutions matters for understanding the sources of environmental ills. While a great deal of effort goes into 'improving decision making' it is of the utmost importance to bear in mind that environmental damage is normally the consequence not of particular moments of decision following prior deliberation by specific individuals, but rather of continuing and endemic institutional patterns of behaviour. As Slapper and Tombs note of corporate crimes more generally, they are not normally the result of the particular intentional acts of particular individuals, but rather 'can be produced by an organisation's structure, its culture, its unquestioned assumptions, its very *modus operandi* . . . ' (Slapper and Tombs 1999: 17).

Responsibility and character

To make these points about the way decisions emerge from institutional routines and patterns of behaviour might be thought to incur the danger of obscuring the assignment of responsibility for choices that lead to public harms. Don't we need there to be a particular discrete decision made by particular agents in order to properly assign responsibility? A virtue of the minuted decision is that it leaves a trail through which one can ascertain who was responsible for a decision and when. Our answer is that the approach we advocate does not deny the need for, nor the possibility of, assigning responsibility for acts; nor does it deny that there may sometimes be virtue in the formal recording and ratification of decisions to achieve that end. However, if we take the model of the formal discrete decision as a paradigm for all decision, the result is actually a loss of a proper assignment of responsibility. Our contention is that one of the reasons for the difficulty in extending the notion of responsibility to include corporate actors has been an overly discrete view of decision making. We will suggest in contrast that the position we have defended here enables us the better to capture a proper understanding of the responsibility that is involved.

The assigning of responsibility in a world of institutional actors is a pressing practical, as well as theoretical, problem. The problems are illustrated in the examples discussed in the last section. Where personal and environmental harms are the result of the behaviour of large-scale complex organisations, there are real difficulties in determining which person, individual or corporate, can be rendered accountable for those harms. The problem is captured nicely in the John Ford film version of Steinbeck's *The Grapes of Wrath*, a novel set against the backdrop of an earlier environmental disaster, the dust bowl. Faced with eviction from his land the farmer, Muley, asks the stark question of accountability, 'who do we shoot?':

Muley's son:	Whose fault is it?
Agent:	You know who owns the land. The Shawnee Land and Cattle Company.
Muley:	And who's the Shawnee Land and Cattle Company?
Agent:	It ain't nobody. It's a company.
Muley's son:	They got a President, ain't they? They got somebody who knows what a shotgun's for, ain't they?
Agent:	Oh son, it ain't his fault, because the bank tells him what to do.
Muley's son:	All right, where's the bank
Agent:	Tulsa. What's the use of pickin' on him? He ain't nothin' but the manager. And he's half-crazy hisself tryin' to keep up with his orders from the East.
Muley:	Then who do we shoot?
Agent:	Brother, I don't know. If I did, I'd tell ya. I just don't know who's to blame.

The question 'who do we shoot?' is asked in a less dramatic form in the legal context of attempting to answer the question 'who should be legally accountable?' in situations where we are dealing with large and complex organisations. Even given an answer to the legal question, there is still the question of ethical responsibility – of whom the different victims, concerned observers and involved agents can call to account for the actions involved. The questions are central to the cases of the *Exxon Valdez* and the *Herald of Free Enterprise* discussed in the last section. What the subsequent official reports in both cases highlighted was the unsatisfactory nature of the ways in which legal accountability was eventually assigned in these cases.

The general theoretical difficulty raised by these cases is normally taken to reside in the way that responsibility appears to be fragmented between different agents within an organisation, and still worse, between different organisations – the problem of many hands. Both knowledge and causal responsibility are dispersed throughout an organisation, rendering it difficult to determine who is to be held accountable. There are a variety of standard responses to the problem (Bovens

1998; Feinberg 1968; Thomson 1980). One response is the notion of corporate responsibility: the idea that the corporate body as a legal person should be rendered accountable. The problem with this response, which is evident from the passage just quoted, is that certain ways of rendering an agent accountable cannot be exercised. One cannot shoot the company. However, other forms of redress may be possible. Another response is to look for some model of personal responsibility, where several different accounts of personal responsibility are in circulation. One is the hierarchical account that is, again, intimated in the passage from *The Grapes of Wrath* – the person we should shoot is the person who sits at the head of an organisation. A second way of assigning personal responsibility is to hold that all those in an organisation should be held collectively responsible. A third way of assigning personal responsibility is to hold individuals accountable to the degree that they were directly involved in the act. These different accounts are often combined in various ways, for example, in the suggestion that we look for both personal and corporate responsibility. However, how they are combined needs some care. For example, part of the problem of developing the notion of corporate responsibility lies in the way in which it has been tied to an overly individualised account of how we understand responsibility. In English law, for example, specific individuals – a director or senior manager – have to be shown to be directly responsible for a death in order for a company to be convicted of manslaughter. One of the reasons for the absence of any convictions for corporate manslaughter against P&O ferries in the case of the sinking of the *Herald of Free Enterprise* was that none of its directors could be singled out for prosecution for the offence (Slapper and Tombs 1999: 31ff and 149ff). Without an individual conviction there could be no corporate conviction.

The question 'who is responsible?' in the context of large and complex institutions is one that has quite understandably dominated discussion of corporate responsibility. However, this question presupposes the answer to a prior question, 'what are they responsible for?' Just as answers to the first question tend to be individualised, so also do answers to this second, prior question. Responsibility is individualised not simply in the sense that it is assigned to specific agents, but also because it is assigned only for particular acts in virtue of particular individual decisions or choices. If one focuses merely on the particular decisions that lead to particular acts, the primary responsibility for action is lost sight of. From that perspective, many acts of environmental damage get assigned to the category of 'accidents' for which no one is held responsible; or if responsibility is assigned, it is to the person with little power who has the misfortune to stand at the bottom end of the organisational chain, and hence is directly involved in the events immediately preceding the 'accident'. However, the problem here lies in taking specific acts and decisions as the primary object of appraisal. As we noted earlier, even in cases of individual decisions we need a wider framework of appraisal that takes historical considerations into account. Choices need to be understood as expressions of character, and responsibility assigned for the kind of characters we become. Again, as noted above, we bear responsibility for the kinds of person we become, for the dispositions of character we develop.

Similar points apply at an institutional level. The question that needs to be asked – and which reports such as those on the *Herald of Free Enterprise* and the *Exxon Valdez* implicitly raise – is 'who is responsible for the character of institutions?' The point of highlighting institutional vices such as sloppiness and complacency in reports is that they highlight the need for responsibility for those vices to be assigned. Asking this question refocuses the way discussions of responsibility for accidents are framed. It is rarely the case that those low down in a hierarchy can be assigned responsibility for the character of an institution, although they may be complicit in the local habits and routines that are constitutive of that character. Such responsibilities in existing hierarchal institutions belong to those who wield power within institutions. Those with power can be held responsible for the character of the institutions of which they are principal agents. Habituated carelessness ought to be no excuse for accidents that are the result, not of a particular decision, but of habitual ways of going on. Those with power bear responsibility for the routines of behaviour that characterise institutions, just as individuals bear responsibility for their own characters. In circumstances such as these, a discrete and individualised model of decision making, far from facilitating the assignment of responsibility, actually frustrates it. Thus our account does not preclude the assignment of responsibility. It rather gives a better understanding of the nature of the responsibility involved.

What makes for good decisions?

Finally we return to our opening question: what makes for good public decisions about the environment? In the first part of this book we criticised a consequentialist answer to this question. We stressed the need for a pluralist understanding of the nature of values and choices, and we suggested that a procedural account of rationality goes some way towards giving us an account of how rational choices can be made in the context of plural and incommensurable values. In the second part of the book we rejected the response to our environmental problems offered by mainstream environmental ethics as an alternative to that of traditional ethical theory. We suggested that mainstream environmental ethics shared a particular misconceived account of the nature of ethical theory with its opponents. In part three of the book we have stressed the need for recognition of the role of narrative and history in our relations to the environments that matter to us. In this concluding chapter we have argued that decisions themselves are not always the discrete events that the standard models of decision making suggest. They can only be properly appraised in terms of historical patterns of choices. Both individuals and institutions develop and express their character through the decisions they make. They can only be properly appraised as such. This account of rational decision making can be understood as a version of the expressive account of rationality that we discussed in chapter 5. Actions are not just instrumental means to an end, but a way of expressing attitudes to people and things. We have suggested that this applies not just to individuals but to institutions also. Our public decisions are ways of forming and expressing the kinds of institutions we inhabit, and they should be assessed as such. Debates about particular decisions and the decision-making process should be understood as part of a larger debate about the character of the communities to which we belong. We have also suggested that the assignment of responsibility likewise requires a wider understanding of the character of institutions. A more historical and more institutional understanding of decisions is required if we are to make sense of the environmental problems that face us. We hope that this book has gone some way towards developing such an understanding.

Bibliography

Agarwal, B. 2001. 'Participatory exclusions, community forestry and gender', *World Development*, 29: 1623–1648.

Allison, G. 1971. *Essence of Decision: Explaining the Cuban Missile Crisis*, Boston: Little, Brown.

Alvares, C. and Billorey, R. 1988. *Damming the Narmada*, Penang, Malaysia: Third World Network.

Anderson, E. 1993. *Value in Ethics and Economics*, Cambridge, MA: Harvard University Press.

Annas, J. 1993. *The Morality of Happiness*, Oxford: Oxford University Press.

Anscombe, G. E. M. 1958. 'Modern moral philosophy', *Philosophy*, 33: 1–19.

Aristotle. 1908. *Metaphysics*, Oxford: Clarendon Press.

Aristotle. 1972. *De Partibus Animalium*, trans. D. Balme, Oxford: Clarendon Press.

Aristotle. 1985. *Nicomachean Ethics*, trans. T. Irwin, Indianapolis: Hackett.

Arneson, R. 1989. 'Equality and equality of opportunity for welfare', *Philosophical Studies*, 55: 77–93.

Arrow, K. 1997. 'Invaluable goods', *Journal of Economic Literature*, 35: 757–765.

Attfield, R. 1987. *A Theory of Value and Obligation*, London: Croom Helm.

Attfield, R. 1994. 'Rehabilitating nature and making nature habitable', in *Philosophy and the Natural Environment*, R. Attfield and A. Belsey (eds), Cambridge: Cambridge University Press.

Barry, B. 1997. 'Sustainability and intergenerational justice', *Theoria*, 45: 43–63.

Barry, J. 1999. *Rethinking Green Politics: Nature, Virtue and Progress*, London: Sage.

Beckerman, W. 1994. '"Sustainable development": is it a useful concept?', *Environmental Values*, 3: 191–209.

Beckerman, W. 2000. 'Review of J. Foster (ed.), *Valuing Nature? Ethics, Economics and the Environment*', *Environmental Values*, 9: 122–124.

Bentham, J. 1970 [1789]. *Introduction to the Principles of Morals and Legislation*, London: Methuen.

Benton, T. 1993. *Natural Relations*, London: Verso.

Bovens, M. 1998. *The Quest for Responsibility*, Cambridge: Cambridge University Press.

Burgess, J., Clark, J. and Harrison, C. 1995. *Valuing Nature: What Lies Behind Responses to Contingent Valuation Surveys?*, London: UCL.

Burnyeat, M. 1980. 'Aristotle on learning to be good', in A. O. Rorty (ed.), *Essays on Aristotle's Ethics*, Berkeley: University of California Press.

Cafaro, P. 2006. *Thoreau's Living Ethics: Walden and the Pursuit of Virtue*, Athens, GA: University of Georgia Press.

Cafaro, P. and Sandler, R. (eds). 2005. *Environmental Virtue Ethics*, Lanham, MD: Rowman & Littlefield Press.

Callicott, J. B. 1980. 'Animal liberation: a triangular affair', *Environmental Ethics*, 2: 311–338.

Callicott, J. B. 1989. *In Defense of the Land Ethic*, Albany, NY: SUNY Press.

Callicott, J. B. 1990. 'The case against moral pluralism', *Environmental Ethics*, 12: 99–124.

Callicott, J. B. 1998. '"Back together again" again', *Environmental Values*, 7: 461–465.

Clifford, S. and King, A. (eds). 1993. *Local Distinctiveness: Place, Particularity and Identity*, London: Common Ground.

Cockburn, A. 1989. 'Trees, cows and cocaine: an interview with Susanna Hecht', *New Left Review*, 173: 34–45.

Cohen, G. A. 1989. 'On the currency of egalitarian justice', *Ethics*, 99: 906–944.

Cohen, J. 1989. 'Deliberation and democratic legitimacy', in *The Good Polity*, A. Hamlin and P. Pettit (eds), Oxford: Blackwell.

Craig, D. 1990. *On the Crofters' Trail*, London: Jonathan Cape.

Crisp, R. and Slote, M. (eds). 1997. *Virtue Ethics*, Oxford: Oxford University Press.

Cronon, W. 1996. 'The trouble with wilderness', in *Uncommon Ground: Toward Reinventing Nature*, W. Cronon (ed.), New York: W. W. Norton.

Daly, H. E. 1995. 'On Wilfred Beckerman's critique of sustainable development', *Environmental Values*, 4, 1: 49–55.

Darwall, S., Gibbard, A. and Railton, P. 1992. 'Towards fin de siècle ethics: some trends', *Philosophical Review*, 101: 115–189.

de-Shalit, A. 1995. *Why Posterity Matters: Environmental Policies and Future Generations*, London: Routledge.

de-Shalit, A. 2000. *The Environment Between Theory and Practice*, Oxford: Oxford University Press.

Dostoevsky, F. 1994. *The Brothers Karamazov*, Oxford: Oxford University Press.

Dryzek, J. 1990. *Discursive Democracy: Politics, Policy and Political Science*, Cambridge: Cambridge University Press.

Dryzek, J. 2000. *Deliberative Democracy and Beyond: Liberals, Critics, Contestations*, Oxford: Oxford University Press.

Dworkin, R. 1977. *Taking Rights Seriously*, London: Duckworth.

Dworkin, R. 1981. 'What is equality? Part I: Equality of welfare; Part II: Equality of resources', *Philosophy and Public Affairs*, 10: 185–246, 283–345.

Eckersley, R. 1998. 'Beyond human racism', *Environmental Values*, 7: 165–182.

Elliot, E. 1995. 'Faking nature', in *Environmental Ethics*, R. Elliot (ed.), Oxford: Oxford University Press.

Elliot, E. 1996. *Faking Nature*, London: Routledge.

Elster, J. 1986. 'The market and the forum: three varieties of political theory', in *Foundations of Social Choice Theory*, J. Elster and A. Hylland (eds), Cambridge: Cambridge University Press.

English Nature. 1993. *Position Statement on Sustainable Development*, Peterborough: English Nature.

Epictetus, 1910. *Enchiridion* XLIII from *The Moral Discourses of Epictetus*, trans. E. Carter, London: Dent.

Eurobarometer. 2000. 'Europeans and biotechnology', *Eurobarometer* 52.1 [online]. Luxemburg: Office for Official Publications of the European Communities. Brussels, European Commission, Research DG. http://europa.eu.int/comm-/research/pdf/euro barometer-en.pdf

FAO. 1997. *The State of the World's Plant Genetic Resources for Food and Agriculture*, Rome: Food and Agricultural Organization of the United Nations.

Feinberg, J. 1968. 'Collective responsibility', *Journal of Philosophy*, 65: 222–251.

Flyvbjerg, B, Bruzelius, N. and Rothengatter, W. 2003. *Megaprojects and Risk: An Anatomy of Ambition*, Cambridge: Cambridge University Press.

Flyvbjerg, B., Holm, M. and Buhl, S. 2002. 'Underestimating costs in public works projects: error or lie?', *Journal of the American Planning Association*, 68: 279–295.

Gamborg, C. and Sandøe, P. 2004. 'Beavers and biodiversity: the ethics of restoration ecology', in *Philosophy and Biodiversity*, M. Oksanen and J. Pietarinen (eds), New York: Cambridge University Press.

Ghiselin, M. 1987. 'Species concepts, individuality, and objectivity', *Biology and Philosophy*, 2: 127–143.

Gillespie, J. and Shepherd, P. 1995. *Establishing Criteria for Identifying Critical Natural Capital in the Terrestrial Environment*, Peterborough: English Nature.

Gleason, H. A. 1927. 'Further views on the succession concept', *Ecology*, 8: 299–326.

Gleason, H. A. 1936. *Memorandum Brooklyn Botanical Garden*, 4.

Goodin, R. 1992. *Green Political Theory*, Cambridge: Polity Press.

Goodpaster, K. 1978. 'On being morally considerable', *Journal of Philosophy*, 75: 308–325.

Guha, R. 1997. 'The authoritarian biologist and the arrogance of anti-humanism', *The Ecologist*, 27, 1: 14–20.

Guha, R. and Martinez-Alier, J. 1997. *Varieties of Environmentalism*, London: Earthscan.

Habermas, J. 1975. *The Legitimation Crisis of Late Capitalism*, trans. T. McCarthy Boston: Beacon Press.

Hargrove, E. 1992. 'Weak anthropocentric intrinsic value', *The Monist*, 75: 183–207.

Hecht, S. 1989. 'Chico Mendes: chronicle of a death foretold', *New Left Review*, 173: 47–55.

Hicks, J. 1981. *Wealth and Welfare*, Oxford: Blackwell.

Higgs, E. 2003. *Nature by Design*, Cambridge, MA: The MIT Press.

Hill Jr, T. E. 1983. 'Ideals of human excellence and preserving natural environments', *Environmental Ethics*, 5(3): 211–224.

HM Government. 1990. *This Common Inheritance*, Cm. 1200, London: HMSO.

HM Government. 1994. *UK Biodiversity Action Plan*, Cm 2428, London: HMSO.

Holland, A. 1994. 'Natural capital', in *Philosophy and the Natural Environment*, A. Belsey and R. Attfield (eds), Cambridge: Cambridge University Press, 169–182.

Holland, A. 1995a. 'The use and abuse of ecological concepts in environmental ethics', *Biodiversity and Conservation*, 4: 812–826. [Reprinted in *Ecologists and Ethical Judgements*, N. Cooper and R. Carling (eds), Chapman & Hall 1996.].

Holland, A. 1995b. 'The assumptions of cost-benefit analysis: a philosopher's view', in *Environmental Valuation: Some New Perspectives*, K. Willis and J. Corkindale (eds), CAB International, 21–38.

Holland, A. 1997. 'Substitutability: or, why strong sustainability is weak and absurdly strong sustainability is not absurd', in *Valuing Nature?*, J. Foster (ed.), London: Routledge.

Holland, A. 2002. 'Are choices trade-offs?', in *Economics, Ethics and Environmental Policy*, D. Bromley and J. Paavola (eds), Oxford: Blackwell, pp. 17–34.

Holland, A. 2004. 'Is nature sacred?', in *Is Nothing Sacred?*, B. Rogers (ed.), London: Routledge.

Holland, A. and Rawles, K. 1994. The *Ethics of Conservation*. Report presented to The Countryside Council for Wales. Thingmount Series, No.1. Lancaster University: Department of Philosophy.

Hume, D. 1978 [1739]. *A Treatise of Human Nature*, Oxford: Clarendon Press.

Hursthouse, R. 2000. *Ethics, Humans and Other Animals*, London: Routledge.

Jacobs, M. 1995. 'Sustainable development, capital substitution and economic humility: a response to Beckerman', *Environmental Values*, 4: 57–68.

Jamieson, D. 1998. 'Animal liberation is an environmental ethic', *Environmental Values*, 7: 41–57.

Johnson, L. E. 1991. *A Morally Deep World*, Cambridge: Cambridge University Press.

Jones, M. 1977. *Finland: Daughter of the Sea*, Folkestone: Dawson.

Kant, I. 1933. *Critique of Pure Reason*, trans. N. Kemp Smith, London: Macmillan.

Kant, I. 1956. *Groundwork of the Metaphysic of Morals*, London: Hutchinson.

Kant, I. 1968. *Lectures on Anthropology*, Berlin: Akademie-Textausgabe.

Kant, I. 1979. *Lectures on Ethics*, London: Methuen.

Katz, E. 1993. 'Artefacts and functions: a note on the value of nature', *Environmental Values*, 2: 223–232.

Katz, E. 1999. 'A pragmatic reconsideration of anthropocentrism', *Environmental Ethics*, 21: 377–390.

Katz, E. and Oechsli, L. 1993. 'Moving beyond anthropocentrism: environmental ethics, development and the Amazon', *Environmental Ethics*, 15: 49–59.

Lawton, J. 1991. 'Are species useful?', *Oikos*, 62: 3–4.

Leopold, A. 1949. *A Sand County Almanac*, New York: Oxford University Press.

Light, A. 2000. 'Ecological restoration and the culture of nature: a pragmatic perspective', in *Restoring Nature: Perspectives from the Social Sciences and Humanities*, P. Gobster and B. Hull (eds), Washington, DC: Island Press.

Light, A. 2002a. 'Contemporary environmental ethics: from metaethics to public philosophy', *Metaphilosophy*, 33: 426–449.

Light, A. 2002b. 'Restoring ecological citizenship', in *Democracy and the Claims of Nature*, B. Minteer and B. P. Taylor (eds), Lanham, MD: Rowman & Littlefield.

Light, A. 2003. 'The case for a practical pluralism', in A. Light and H. Rolston III (eds), *Environmental Ethics: An Anthology*, Oxford: Blackwell.

Light, A. forthcoming. '"Faking nature" revisited', in *The Beauty Around Us: Environmental Aesthetics in the Scenic Landscape and Beyond*, D. Michelfelder and B. Wilcox (eds), Albany, NY: SUNY Press.

Lukes, S. 1997. 'Comparing the incomparable: trade-offs and sacrifices', in *Incommensurability, Incomparability and Practical Reason*, R. Chang (ed.), Cambridge, MA: Harvard University Press, pp. 184–195.

Lund, K. 2006. 'Finding place in nature: "intellectual" and local knowledge in a Spanish natural park', *Conservation and Society*, 3: 371–387.

Lynch, T and Wells, D. 1998. 'Non-anthropocentrism? A killing objection', *Environmental Values*, 7: 151–163.

Mabey, R. 1980. *The Common Ground*, London: Hutchinson.

MacIntyre, A. 1986. *After Virtue*, London: Duckworth.

Mackie, J. 1977. *Ethics*, Harmondsworth: Penguin.

Macleod, C. 1983. *Collected Essays*, Oxford: Clarendon Press.

Macnaghten, P. 2004. 'Animals in their nature: a case study of public attitudes to animals, genetic modification and "nature"', *Sociology*, 38: 533–551.

Mahalia, B. 1994. 'Letter from a tribal village', *Lokayan Bulletin*, 11.2/3, Sept–Dec.

Marmot, M. 2004. *The Status Syndrome*, London: Bloomsbury.

Martinez-Alier, J. 1997. 'The merchandising of biodiversity', in *Justice, Property and the Environment: Social and Legal Perspectives*, T. Hayward and J. O'Neill (eds), Aldershot: Avebury.

Martinez-Alier, J. 2002. *The Environmentalism of the Poor*, Cheltenham: Edward Elgar.

Martinez-Alier, J., Munda, G. and O'Neill, J. 1999. 'Commensurability and compensability in ecological economics', in *Valuation and Environment: Principles and Practices*, C. Spash and M. O'Connor (eds), Aldershot: Edward Elgar.

Marx, K. 1974 [1844]. *Economic and Philosophical Manuscripts*, in *Early Writings*, L. Colletti (ed.), Harmondsworth: Penguin.

Mayr, E. 1987. 'The ontological status of species: scientific progress and philosophical terminology', *Biology and Philosophy*, 2: 145–166.

McCully, P. 1996. *Silenced Rivers: The Ecology and Politics of Large Dams*, London: Zed Books.

McKibben, B. 1990. *The End of Nature*, New York: Viking.

McNaughton, D. 1988. *Moral Vision*, Oxford: Blackwell.

Mill, J. S. 1994 [1848]. *Principles of Political Economy*, Oxford: Oxford University Press.

Mill, J. S. 1962 [1861]. *Utilitarianism*, in M. Warnock (ed.), *Utilitarianism*, London: Collins.

Mill, J. S. 1874. 'Nature', in *Three Essays on Religion*, London: Longmans.

Mill, J. S. 1884. *A System of Logic*, New York: Harper and Brothers.

Miller, D. 1992. 'Deliberative democracy and social choice', *Political Studies*, 40: 54–67.

Miller, D. 1997. 'Equality and justice', *Ratio* (new series), 10: 222–237.

Montaigne, M. 1958. 'On Friendship', in *Essays*, trans. J. M. Cohen, Harmondsworth: Penguin.

Moore, G. E. 1922. 'The conception of intrinsic value', in *Philosophical Studies*, London: Routledge & Kegan Paul.

Naess, A. 1973. 'The shallow and the deep, long range ecology movement', *Inquiry*, 16: 95–100.

Naess, A. 1984. 'A defence of the deep ecology movement', *Environmental Ethics*, 6: 265–270.

Norman, R. 1997. 'The social basis of equality', *Ratio* (new series), 10: 238–252.

Norman, R. 2004. 'Nature, science, and the sacred', in *Is Nothing Sacred?*, B. Rogers (ed.), London: Routledge.

North, R. 1990. 'Why man can never kill nature', *Independent*, 1 February.

Norton, B. G. 1984. 'Environmental ethics and weak anthropocentrism', *Environmental Ethics*, 6: 131–148.

Norton, B. G. 1987. *Why Preserve Natural Variety?*, Princeton: Princeton University Press.

Norton, B. G. 2005. *Sustainability: A New Philosophy of Adaptive Ecosystem Management*, Chicago: University of Chicago Press.

Nozick, R. 1974. *Anarchy, State and Utopia*, Oxford: Blackwell.

Nuffield Council. 1999. *Genetically Modified Crops: The Ethical and Social Issues*, London: Nuffield Council on Bioethics.

O'Connor, M. (ed.). 1994. *Is Capitalism Sustainable?*, New York: The Guilford Press.

O'Connor, M. and Muir, E. 1995. 'Endowment effects in competitive general equilibrium: a primer for Paretian policy analysts', *Journal of Income Distribution*, 5: 145–175.

Olwig, K. 1996. 'Reinventing common nature: Yosemite and Mt. Rushmore – a meandering tale of a double nature', in *Uncommon Ground: Toward Reinventing Nature*, W. Cronon (ed.) New York: W. W. Norton.

O'Neill, J. 1993. *Ecology, Policy and Politics: Human Well-Being and the Natural World*, London: Routledge.

O'Neill, J. 1995. 'Polity, economy, neutrality', *Political Studies*, 43: 414–431.

O'Neill, J. 1997a. 'Cantona and Aquinas on good and evil', *Journal of Applied Philosophy*, 14: 97–106.

O'Neill, J. 1997b. 'Time, narrative and environmental politics', in *New Perspectives in Environmental Politics*, R. Gottlieb (ed.), London: Routledge.

O'Neill, J. 2001. 'Chekov and the egalitarian', *Ratio* (new series), 14:165–170.

O'Neill, J. 2007. *Markets, Deliberation and Environment*, London: Routledge.

Orwell, G. 1966. *Homage to Catalonia*, Harmondsworth: Penguin.

Ouderkirk, W. 1998. 'Mindful of the earth: a bibliographical essay on environmental philosophy', *The Centennial Review*, 42: 353–392.

Palmer, C. 2003. 'An overview of environmental ethics', in *Environmental Ethics: An Anthology*, A. Light and H. Rolston III (eds), Malden, MA: Blackwell Publishers.

Parfit, D. 1984. *Reasons and Persons*, Oxford: Oxford University Press.

Parfit, D. 1997. 'Equality and priority', *Ratio* (new series), 10: 202–221.

Pearce, D. 1993. *Economic Values and the Natural World*, London: Earthscan.

Pearce, D. and Moran, D. 1995. *The Economic Value of Biodiversity*, London: Earthscan.

Pearce, D., Markandya A. and Barbier, E. 1989. *Blueprint for a Green Economy*, London: Earthscan.

Pigou, A. 1952. *The Economics of Welfare* (4th edn), London: Macmillan.

Plumwood, V. 1993. *Feminism and the Mastery of Nature*, London: Routledge.

Plumwood, V. 1998. 'Intentional recognition and reductive rationality: a response to John Andrews', *Environmental Values*, 7, 4: 397–421.

Rawls, J. 1972. *A Theory of Justice*, Oxford: Oxford University Press.

Raz, J. 1986. *The Morality of Freedom*, Oxford: Clarendon Press.

Regan, T. 1988. *The Case for Animal Rights*, London: Routledge.

Ridley, M. 1985. *The Problems of Evolution*, Oxford: Oxford University Press.

Ritchie, D. 1916. *Natural Rights*, 3rd edn. London: Allen & Unwin.

Robbins, L. 1938. 'Interpersonal comparisons of utility', *The Economic Journal*, 48: 635–641.

Rolston III, H. 1988. *Environmental Ethics*, Philadelphia: Temple University Press.

Rolston III, H. 1990. 'Duties to ecosystems', in *Companion to a Sand County Almanac*, J. Baird Callicott (ed.), Wisconsin: University of Wisconsin Press.

Ross, D. 1930. *The Right and the Good*, Oxford: Oxford University Press.

Routley, R. and V. 1980. 'Human chauvinism and environmental ethics', in *Environmental Philosophy*, D. Mannison, M. McRobbie and R. Routley (eds), Canberra: Australian National University.

Runte, A. 1987. *National Parks: The American Experience*, 2nd edn, Lincoln: University of Nebraska Press.

Russell, D. 1998. 'Forestry and the art of frying small fish', *Environmental Values*, 7: 281–290.

Samuels, W. 1981. 'Welfare economics, power and property', in *Law and Economics: An International Perspective*, W. Samuels and A. Schmid (eds), Boston: Martin Nijhoff.

Sandel, M. 1982. *Liberalism and the Limits of Justice*, Cambridge: Cambridge University Press.

Sen, A. 1980. 'Equality of what?', in *Tanner Lectures on Human Values,* S. McMurrin (ed.), Cambridge: Cambridge University Press.

Sen, A. 1987. *On Ethics and Economics*, Oxford: Blackwell.

Sen, A. 1997. *On Economic Inequality*, 2nd edn, Oxford: Clarendon Press.

Shiva, V. 1992. 'Recovering the real meaning of sustainability', in *The Environment in Question*, D. Cooper and J. Palmer (eds), London: Routledge.

Siipi, H. 2003. 'Artefacts and living artefacts', *Environmental Values*, 12: 413–430.

Simon, H. 1979. 'From substantive to procedural rationality', in *Philosophy and Economic Theory*, F. Hahn and M. Hollis (eds), Oxford: Oxford University Press.

Singer, P. 1976. *Animal Liberation*, London: Jonathan Cape.

Singer, P. 1986. 'All animals are equal', in *Applied Ethics*, P. Singer (ed.), Oxford: Oxford University Press.

Slapper, G. and Tombs, S. 1999. *Corporate Crime*, Harlow: Longman.

Smith, G. 2003. *Deliberative Democracy and the Environment*, London: Routledge.

Sober, E. 1986. 'Philosophical problems for environmentalism', in *The Preservation of Species: The Value of Biological Diversity*, Princeton: Princeton University Press.

Solow, R. M. 1974. 'The economics of resources or the resources of economics', *American Economic Review*, 64: 1–14.

Spash, C. and Hanley, N. 1994. 'Preferences, information and biodiversity preservation', *Ecological Economics*, 12: 191–208.

Steiner, H. 1994. *An Essay on Rights*, Oxford: Blackwell.

Stevenson, C. L. 1944. *Ethics and Language*, New Haven: Yale University Press.

Stone, C. 1988. 'Moral pluralism and the course of environmental ethics', *Environmental Ethics*, 10: 139–154.

Suomin, J. 1979. 'The grain immigrant flora of Finland', *Acta Botannica Fennica*, 111: 1–108.

Sylvan, R. 2003 [1973]. 'Is there a need for a new, an environmental, ethic?', in *Environmental Ethics: An Anthology*, A. Light and H. Rolston III (eds), Malden, MA: Blackwell Publishers.

Sylvan, R. 1998. 'Mucking with nature', in *Applied Ethics in a Troubled World*, E. Morscher (ed.), Dordrecht: Kluwer.

Tawney, R. 1964. *Equality*, London: Unwin.

Taylor, P. 1986. *Respect for Nature*, Princeton: Princeton University Press.

Temkin, L. 1993. *Inequality*, Oxford: Oxford University Press.

Thomson, D. 1980. 'Moral responsibility and public officials: the problem of many hands', *American Political Science Review*, 74: 905–916.

US Office of Technology Assessment. 1987. *Technologies to Maintain Biological Diversity*, Washington DC: Government Printing Office.

Varner, G. 1998. *In Nature's Interests*, Oxford: Oxford University Press.

Vatn, A. and Bromley, D. 1994. 'Choices without prices without apologies', *Journal of Environmental Economics and Management*, 26: 129–148.

Walsh, V. 1970. *Introduction to Contemporary Microeconomics*, New York, McGraw-Hill.

Wenz, P. 1993. 'Minimal, moderate, and extreme moral pluralism', *Environmental Ethics*, 15: 61–74.

Wenz, P. 2001. *Environmental Ethics Today*, Oxford: Oxford University Press.

Wiggins, D. 1991. 'The claims of need', in *Needs, Values, Truth*, 2nd edn, Oxford: Blackwell.

Wiggins, D. 2000. 'Nature, respect for nature, and the human scale of values', *Proceedings of the Aristotelian Society*, 100: 1–32.

Williams, B. 1973. 'A critique of utilitarianism', in *Utilitarianism For and Against*, J. J. C. Smart and B. Williams, Cambridge: Cambridge University Press.

Williams, B. 1985. *Ethics and the Limits of Philosophy*, London: Collins.

Williams, R. 1976. *Keywords*, London: Fontana.

Wilson, R. 1999. *Species: New Interdisciplinary Essays*, Cambridge, MA: MIT Press.

Wood, P. 1997. 'Biodiversity as the source of biological resources: a new look at biodiversity values', *Environmental Values*, 6: 251–268.

World Commission on Environment and Development. 1987. *Our Common Future*, London: Oxford University Press.

Worster, D. 1977. *Nature's Economy*, Cambridge: Cambridge University Press.

Wright, G. H. von. 1963. *The Varieties of Goodness*, London: Routledge & Kegan Paul.

Name index

Agarwal, B. 49
Allison, G. 207
Alvares, C. 51
Anderson, E. 84
Annas, J. 47, 82
Anscombe, G.E.M., 41
Aristotle 41, 42, 106, 109, 209
Arneson, R. 67
Attfield, R. 100, 102–4, 119, 144–145

Barry, B. 67
Barry, J. 41
Beckerman, W. 186–8
Bentham, J. 15–16, 17, 18, 19, 26–7, 52, 70–2,
    74, 75, 96–7
Benton, T. 38–9
Billorey, R. 51
Bovens, M. 213–14
Brundtland, G.H. 183–4, 195
Burgess, J. 78
Burnyeat, M. 209

Cafaro, P. 41–42
Callicott, J. B. 70, 100, 105–6
Cameron, J. 37
Clements, F. 136
Clifford, S. 3, 176
Cockburn, A. 181
Cohen, J. 204
Cohen, G. A. 67
Craig, D. 157
Crisp, R, 41
Cronon, W. 132

Daly, H. E. 187–8, 190–1
Darius 77–8
Darwall, S. 112
de-Shalit, A. 67
Dostoevsky, F. 32
Dryzek, J. 204
Dworkin, R. 36, 67, 73

Eckersley, R. 181
Elliot, R. 139–46, 149, 163
Elster, J. 204
Epictetus, 155

Feinberg, J. 214
Flyvbjerg, B. 52, 210

Gamborg, C. 147
Ghiselin, M. 178
Gillespie, J. 174
Gleason, H. A. 138
Goodin, R. 125, 148, 163
Goodpaster, K. 93
Guha, R. 49, 69

Habermas, J. 205
Hanley, N. 73
Hargrove, E. 116
Hecht, S. 181
Heraclitus, 42
Hicks, J. 29–30, 55, 56, 57, 58, 59
Higgs, E. 143
Hobbes, T. 131
Hume, D. 126–8, 130, 131
Hursthouse, R. 41

Jacobs, M. 67, 187, 199–200
Jamieson, D. 106
Johnson, L. E. 138
Jones, M. 161

Kaldor, N. 29–30, 55, 56, 57, 58, 59
Kant, I. 33–5, 38, 94–6, 98, 104, 109, 115, 134,
    204–5
Katz, E. 116, 134–7, 141–2, 146, 180–1
King, A. 3, 176

Lawton, J. 137
Leibniz, G. 117
Leopold, A. 91, 100, 104–5, 138, 143, 160

Subject index